STUDIES IN LOGIC

BY MEMBERS

OF THE

JOHN HOPKINS UNIVERSITY

Elibron Classics
www.elibron.com

Elibron Classics series.

ISBN 1-4021-9966-X (paperback)
ISBN 1-4212-9988-7 (hardcover)

This Elibron Classics Replica Edition is an unabridged facsimile of the edition published in 1883 by Little, Brown, and Company, Boston.

STUDIES IN LOGIC.

BY MEMBERS

OF THE

JOHNS HOPKINS UNIVERSITY.

LITTLE, BROWN, AND COMPANY.

1883

PREFACE.

These papers, the work of my students, have been so instructive to me, that I have asked and obtained permission to publish them in one volume.

Two of them, the contributions of Miss Ladd (now Mrs. Fabian Franklin) and of Mr. Mitchell, present new developments of the logical algebra of Boole. Miss Ladd's article may serve, for those who are unacquainted with Boole's "Laws of Thought," as an introduction to the most wonderful and fecund discovery of modern logic. The followers of Boole have altered their master's notation mainly in three respects.

1. A series of writers, — Jevons, in 1864; Peirce, in 1867; Grassman, in 1872; Schröder, in 1877; and McColl in 1877, — successively and independently declared in favor of using the sign of addition to unite different terms into one aggregate, whether they be mutually exclusive or not. Thus, we now write

European + Republican,

to stand for all Europeans and Republicans taken

together, without intending to count twice over the European Republicans. Boole and Venn (his sole living defender) would insist upon our writing

European + Non-European Republican,

or

Non-Republican European + Republican.

The two new authors both side with the majority in this respect.

2. Mr. McColl and I find it to be absolutely necessary to add some new sign to express *existence;* for Boole's notation is only capable of representing that some description of thing does *not* exist, and cannot say that anything *does* exist. Besides that, the sign of equality, used by Boole in the desire to assimilate the algebra of logic to that of number, really expresses, as De Morgan showed forty years ago, a complex relation. To say that

African = Negro

implies two things, that every African is a Negro, and that every Negro is an African. For these reasons, Mr. McColl and I make use of signs of inclusion and of non-inclusion. Thus, I write

Griffin $-\!\!<$ breathing fire

to mean that every griffin (if there be such a creature) breathes fire; that is, no griffin not breathing fire exists; and I write

Animal $\overline{-\!\!<}$ Aquatic,

to mean that some animals are not aquatic, or

that a non-aquatic animal does exist. Mr. McColl's notation is not essentially different.

Miss Ladd and Mr. Mitchell also use two signs expressive of simple relations involving existence and non-existence; but in their choice of these relations they diverge both from McColl and me, and from one another. In fact, of the eight simple relations of terms signalized by De Morgan, Mr. McColl and I have chosen two, Miss Ladd two others, Mr. Mitchell a fifth and sixth. The logical world is thus in a situation to weigh the advantages and disadvantages of the different systems.

3. The third important modification of Boole's original notation consists in the introduction of new signs, so as to adapt it to the expression of relative terms. This branch of logic which has been studied by Leslie Ellis, De Morgan, Joseph John Murphy, Alexander MacFarlane, and myself, presents a rich and new field for investigation. A part of Mr. Mitchell's paper touches this subject in an exceedingly interesting way.

The method of using the Boolian calculus — already greatly simplified by Schröder and by McColl — receives still further improvements at the hands both of Miss Ladd and Mr. Mitchell, and it is surprising to see with what facility their methods yield solutions of problems more intricate and difficult than any that have hitherto been proposed.

The volume contains two other papers relating to deductive logic. In one of these Mr. Gilman develops those rules for the combination of relative numbers of which the general principles of probabilities are special cases. In the other, Dr. Marquand shows how a counting machine, on a binary system of numeration, will exhibit De Morgan's eight modes of universal syllogism.

There are, besides, two papers upon inductive logic. In the first, Dr. Marquand explains the deeply interesting views of the Epicureans, known to us mainly through the work of Philodemus, *περὶ σημείων καὶ σημειώσεων*, which exists in a fragmentary state in a Herculaneum papyrus.

The other paper is one which, at the desire of my students, I have contributed to the collection. It contains a statement of what appears to me to be the true theory of the inductive process, and the correct maxims for the performance of it. I hope that the thoughts that a long study has suggested to me may be found not altogether useless to those who occupy themselves with the application of this kind of reasoning.

I have to thank the Trustees of the Johns Hopkins University, for a very liberal contribution toward the expenses of this publication.

C. S. PEIRCE.

BALTIMORE, Dec. 12, 1882.

CONTENTS.

THE LOGIC OF THE EPICUREANS.

By Allan Marquand.

When we think of the Epicureans we picture a friendly brotherhood in a garden, soothing each other's fears, and seeking to realize a life of undisturbed peace and happiness. It was easy, and to their opponents it became natural, to suppose that the Epicureans did not concern themselves with logic; and if we expect to find in their writings a highly developed formal logic, as that of the Stoics, our search will be in vain. But if we examine the letters of Epicurus, the poem of Lucretius, and the treatise of Philodemus[1] with a view to discovering the Epicurean mode of thought, we find a logic which outweighs in value that of their Stoic rivals. This logic is interesting to us, not only because it is the key to that school of Greek Philosophy which outlasted every other, but because a similar logic controls a powerful school of English thought.

The logic of Epicurus, like that of J. S. Mill, in opposition to conceptualism, attempts to place philosophy upon an empirical basis. Words with Epicurus are signs of things, and not, as with the Stoics, of our ideas of

[1] Gomperz: Herkulanische Studien I. Leipzig, 1865. Bahnsch: Des Epicureers Philodemus Schrift Περὶ σημείων καὶ σημειώσεων. Eine Darlegung ihres Gedankengehalts. Lyck, 1879.

things.[1] There are, therefore, two methods of inquiry: One seeks for the meanings of words; the other, for a knowledge of things. The former is regarded as a preliminary process; the latter, the only true and necessary way of reaching a philosophy of the universe.

All our knowledge is to be brought to the test of sensation, pre-notion, and feeling.[2] By these we do not understand three ultimate sources of knowledge. Democritus[3] held to only one source, viz., Feeling; and Epicurus, who inherited his system, implicitly does the same. But each of these modes of feeling has its distinguishing characteristic, and may be used to test the validity of our knowledge. It is the peculiarity of sensation to reveal to us the external world. Sensation[4] reasons not, remembers not; it adds nothing, it subtracts nothing. What it gives is a simple, self-evident, and true account of the external world. Its testimony is beyond criticism. Error arises after the data of sensation become involved in the operations of intellect. If we should compare this first test of truth with Hume's "impressions," the second test, pre-notion, would correspond with Hume's "ideas." Pre-notions[5] were copies of sensations in a generalized or typical form, arising from a repetition of similar sensations. Thus the belief in the gods[6] was referred to the clear pre-notions of them. Single effluxes from such refined beings could have no effect upon the senses, but repeated effluxes from deities sufficiently similar produce in our minds the general notion of a god.[7] In the same

[1] The hypothesis of λεκτά, or of immaterial notions, was a conceptualistic inconsistency on the part of the Stoics. The Epicureans and the more consistent empiricists among the Stoics rejected them. Sextus Empiricus, Math. viii. 258.

[2] Diogenes Laertius, x. 31.

[3] Sextus: Math., vii. 140.

[4] D. L., x. 31.

[5] D. L., x. 33.

[6] D. L., x. 123, 124.

[7] Cicero: De Nat. Deor., i. 49; D. L,. x. 139.

manner, but through the senses, the continued observation of horses or oxen produce in us general notions, to which we may refer a doubt concerning the nature of the animal that moves before us.

The third criterion, Feeling (in the limited sense), was the ultimate test for ethical maxims. The elementary forms are the feeling of pleasure and the feeling of pain. A fourth criterion was added, viz., The Imaginative representations of the intellect. Its use is by no means clear.

Upon this foundation rises the structure of Epicurean logic. When we leave the clear evidence of sense we pass into the region of opinion, away from the stronghold of truth to the region where error is ever struggling for the mastery of our minds. A true opinion [1] is characterized as one for which there is evidence in favor or none against; a false opinion, one for which there is no evidence in favor or some against. The processes by which we pass to the more general and complex forms of knowledge are four: Observation, Analogy, Resemblance, Synthesis.[2] By Observation, we come into contact with the data of the senses; by Analogy, we may not only enlarge and diminish our perceptions, as we do in conceiving a Cyclops or a Pygmy, but also extend to the unperceived the attributes of our perceptions, as we do in assigning properties to atoms, the soul, and the gods; by Resemblance, we know the appearance of Socrates from having seen his statue; by Synthesis, we combine sensations, as when we conceive of a Centaur.

As a matter of fact, Epicurus regards only two processes, — Observation and Analogy. Our knowledge, then,

[1] D. L., x. 34, 51. Sextus: Math., vii. 211.

[2] D. L., x. 32. The Stoics held a similar view; see D. L., vii. 52.

consists of two parts:[1] (1) The observed, or phenomena clear and distinct to consciousness; and (2) The unobserved,[2] consisting of phenomena which are yet to be observed, and of hidden causes which lie forever beyond our observation. The function of logic[3] consists in inference from the observed to the unobserved. This was called a sign-inference. According to Epicurus there are two methods[4] of making such an inference; one resulting in a single explanation, the other in many explanations. The former may be illustrated by the argument,—Motion is a sign of a void. Here the void is regarded as the only explanation to be given of motion. In other cases many explanations are found equally in harmony with our experience. All celestial phenomena belong to this class. That explanation which alone represents the true cause of such a phenomenon being unknown, we must be content to admit many explanations as equally probable. Thus thunder[5] is explained by supposing either that winds are whirling in the cavities of the clouds, or that some great fire is crackling as it is fanned by the winds, or that the clouds are being torn asunder or are rubbing against each other as they become crystallized. In thus connecting celestial and terrestrial phenomena, Epicurus aimed only to exclude supernaturalistic explanations. This done, he was satisfied.

In the garden at Athens this logic took root and grew; and by the time that Cicero visited Greece and sat at the feet of Zeno,[6] he may have listened to that great repre-

[1] Philodemus: Rhet., lib. iv., i. col. xix.

[2] That is, *τὸ προσμένον καὶ τὸ ἄδηλον*, D. L., x. 38.

[3] D. L., x. 32. *ὅθεν καὶ περὶ τῶν ἀδήλων ἀπὸ τῶν φαινομένων Χρὴ σημειοῦσθαι.*

[4] Ibid., x. 86, 87.

[5] Ibid., x. 100. Cf. Lucretius, lib. vi. 95–158.

[6] See Zeller's Stoics, Epicureans, and Sceptics. London, 1880, p. 412, *n.* 3.

sentative of the Epicurean School discussing such questions[1] as,—How may we pass from the known to the unknown? Must we examine every instance before we make an induction? Must the phenomenon taken as a sign be identical with the thing signified? Or, if differences be admitted, upon what grounds may an inductive inference be made? And, Are we not always liable to be thwarted by the existence of exceptional cases?—But such questions had no interest for Cicero. He was too much an orator and rhetorician to recognize the force of the Epicurean opposition to dialectic. The Epicurean logic[2] to him was barren and empty. It made little of definition; it said nothing of division; it erected no syllogistic forms; it did not direct us how to solve fallacies and detect ambiguities. And how many have been the historians of philosophy who have assigned almost a blank page to Epicurean logic!

With a supreme confidence in the truth of sensation and the validity of induction the Epicureans stood in conflict with the other schools of Greek philosophy. The Stoics, treating all affirmation from the standpoint of the hypothetical proposition, acknowledged the validity of such inductions only as could be submitted to the *modus tollens*. The Sceptics denied the validity of induction altogether. Induction was treated as a sign-inference, and a controversy appears to have arisen concerning the nature of signs, as well as concerning the mode and validity of the inference. The Stoics divided signs into *suggestive* and *indicative*.[3] By means of a suggestive sign we recall some previously associated fact: as from smoke we infer fire. By indicative signs we infer something otherwise unknown: thus motions of

[1] Philodemus περι σημειῶν, col. xix.–xx. [2] Cicero: De Fin., i. 7, 22.
[3] See Prandtl's Ges. d. Log., i. 458.

the body are signs of the soul. Objectively a sign was viewed as the antecedent of a valid conditional proposition, implying a consequent. Subjectively, it was a thought, mediating in some way between things on the one hand, and names and propositions on the other. The Epicureans looked upon a sign as a phenomenon, from whose characters we might infer the characters of other phenomena under conditions of existence sufficiently similar. The sign was to them an object of sense. In considering the variety of signs, the Epicureans appear to have admitted three kinds; but only two are defined in the treatise of Philodemus.[1] A general sign is described as a phenomenon which can exist whether the thing signified exists or not, or has a particular character or not. A particular sign is a phenomenon which can exist only on the condition that the thing signified actually exists. The relation between sign and thing signified in the former case is resemblance; in the latter, it is invariable sequence or causality. The Stoics, in developing the sign-inference, inquired, How may we pass from the antecedent to the consequent of a conditional proposition? They replied, A true sign exists only when both antecedent and consequent are true.[2] As a test, we should be able to contrapose the proposition, and see that from the negative of the consequent the negative of the antecedent followed. Only those propositions which admitted of contraposition were allowed to be treated as hypothetical.[3]

On this propositional ground, therefore, the Epicurean must meet his opponent. This he does by observing that general propositions are obtained neither by contraposition nor by syllogism, nor in any other way than

[1] Philod., loc. cit., col. xiv. [2] Sextus : Math., viii. 256.
[3] Cicero: De Fato, 6, 12; 8, 15.

by induction.[1] The contraposed forms, being general propositions, rest also on induction. Hence, if the inductive mode of reasoning be uncertain, the same degree of uncertainty attaches to propositions in the contraposed form.[2] The Stoics, therefore, in neglecting induction, were accused of surrendering the vouchers by which alone their generalizations could be established.[3] In like manner they were accused of hasty generalization, of inaccurate reasoning, of adopting myths, of being rhetoricians rather than investigators of Nature. Into the truth of these accusations we need not inquire. It is enough that they cleared the way for the Epicureans to set up a theory of induction.

The first question which Zeno sought to answer was, "Is it necessary that we should examine every case of a phenomenon, or only a certain number of cases?"[4] Stoics and Sceptics answered, The former is impossible, and the latter leaves induction insecure. But Zeno replied: "It is neither necessary to take into consideration every phenomenon in our experience, nor a few cases at random; but taking *many* and *various* phenomena of the *same general kind*, and having obtained, both from our observation and that of others, the properties that are common to each individual, from these cases may we pass to the rest."[5] Instances taken from a class and exhibiting some invariable properties are made the basis of the inductive inference. A certain amount of variation in the properties is not excluded. Thus from the fact that the men in our region of country are short-lived, we may not infer that the inhabitants of Mt. Athos are short-lived also; for "men in our experience are seen to vary considerably in respect to length or brevity of life."[6]

[1] Philod., loc. cit., col. xvii.
[2] Ibid., col. ix.
[3] Ibid., col. xxix.
[4] Ibid., col. xix. 13–15.
[5] Ibid., col. xx. 30–col. xxi. 3.
[6] Ibid., col. xvii. 18–22.

Within limits, then, we may allow for variation due to the influence of climate, food, and other physical conditions; but our inference should not greatly exceed the limits of our experience. But, in spite of variations, there are properties which in our experience are universal. Men are found to be liable to disease and old age and death; they die when their heads are cut off, or their hearts extracted; they cannot pass through solid bodies. By induction we infer that these characteristics belong to men wherever they may be found, and it is absurd to speak of men under similar conditions as not susceptible to disease or death, or as having the ability to pass through iron as we pass through the air.[1]

The Epicurean looks out upon Nature as already divided and subdivided into classes, each class being closely related to other classes. The inductive inference proceeds from class to class, not in a hap-hazard way, but from one class to that which resembles it most closely.[2] In case the classes are identical, there is no distinction of known and unknown; and hence, properly speaking, no inductive inference.[3] In case the classes are widely different, the inference is insecure. But within a certain range of resemblance we may rely as confidently upon an inductive inference as we do upon the evidence of sense.[4]

In speaking of the common or essential characters, the basis of induction, it was usual to connect them with the subject of discourse by the words ᾗ, *καθό*, or *παρό*. These words may be taken in four senses:[5] (1) The properties may be regarded as necessary consequences; so we may say of a man that he is necessarily corporeal and liable to disease and death. (2) Or as essential to the conception or definition of the subject. This is what is con-

[1] Philod., loc. cit., col. xxi.

[2] Ibid., col. xviii. 20; col. xxviii. 25–29.

[3] Ibid., col. vi. 8–10.

[4] Ibid., Frag. 2, 5–6.

[5] Ibid., col. xxxiii. 33–col. xxxiv. 34.

veyed in the expression, "Body *as* body has weight and resistance; man *as* man is a rational animal." (3) That certain properties are always concomitant. (4) The fourth sense, lost in the lacunæ, appears from the following examples to involve degree or proportion: "The sword cuts *as* it has been sharpened; atoms are imperishable in so far as they are perfect; bodies gravitate in proportion to their weight."

Zeno's theory of induction may be formulated in the following Canons:—

CANON I.—If we examine many and various instances of a phenomenon, and find some character common to them all, and no instance appears to the contrary, this character may be transferred to other unexamined individuals of the same class, and even to other closely related classes.

CANON II.—If in our experience a given character is found to vary, a corresponding amount of variation may be inferred to exist beyond our experience.

The most important objection made to this theory was, that phenomena exist in our experience exhibiting peculiar and exceptional characters, and that other exceptions might exist beyond our experience to vitiate any induction we may make. The following examples are given:[1] The loadstone has the peculiar property of attracting iron; amber, of attracting bran; the square number 4 × 4, of having its perimeter equal to its area. Exceptional characters are found in the Alexandrian anvil-headed dwarf, the Epidaurian hermaphrodite, the Cretan giant, the pygmies in Achoris. The sun and moon also are unique; so are time and the soul. Admitting such exceptional phenomena, the Epicurean replies, that the belief that a similar state of things exists beyond our experience can

[1] Philod., loc. cit., col. i., ii.

be justified only inductively.[1] And exceptional phenomena must be viewed not as closely resembling, but as being widely different from, other phenomena. Inductions concerning loadstones must be confined to loadstones, and not extended to other kinds of stones. Each class of exceptional phenomena offered a new field for induction, and hence could be said to strengthen and not to weaken the inductive argument.[2]

The correctness of all inductions could be tested by the rule of Epicurus for the truth of opinion in general. An induction is true, when all known instances are in its favor, or none against; it is false, when no instances are in its favor, or some against. When the instances are partly one way and partly another, we cannot reach universal conclusions, but only such as are probable.[3]

This theory of induction was completed by a consideration of fallacies, summarized in a work called the "Demetriac."[4] These consisted in —

1. Failing to see in what cases contraposition is applicable.

2. Failing to see that we should make inductions not in a hap-hazard way, but from properties which resemble each other very closely.

3. Failing to see that exceptional phenomena are in no way at variance with the inductive inference, but on the other hand add to its force.

4. Failing to observe that we infer from the known to the unknown, only when all the evidence is in favor and no shadow of evidence appears to the contrary.

5. The failure to perceive that general propositions are derived not by contraposition, but by induction.

When we compare the work of Zeno with that of

[1] Philod., loc. cit., col. xxv.
[2] Ibid., col. xxiv. 10–col. xxv. 2.
[3] Ibid., col. xxv. 31–34.
[4] Ibid., col. xxviii. 13–col. xxix. 24.

Epicurus, an important logical difference is brought to view. Both are occupied with the sign-inference, and look upon inference as proceeding from the known to the unknown. Epicurus, however, sought only by means of hypothesis to explain special phenomena of Nature. Zeno investigated generalizations from experience, with a view to discovering the *validity* of extending them beyond our experience. This resulted in a theory of induction, which, so far as we know, Epicurus did not possess. In the system of Aristotle, induction was viewed through the forms of syllogism, and its empirical foundation was not held in view. The Epicureans, therefore, were as much opposed to the Aristotelian induction, as they were to the Aristotelian syllogism. It was Zeno the Epicurean who made the first attempt to justify the validity of induction. The record of this attempt will give the treatise of Philodemus a permanent value in the history of inductive logic.

It is refreshing to see the formalistic and rhetorical atmosphere which had surrounded the subject of logic breaking away, and an honest attempt being made to justify the premises of syllogism. As yet, this had not been done by all the moods of the philosophers.

It is also interesting to find in the ancient world a theory of induction which rests upon observation, suggests experiment, assumes the uniformity of Nature, and allows for the variation of characters.

A MACHINE FOR PRODUCING SYLLOGISTIC VARIATIONS.

By Allan Marquand.

From any syllogism a number of logical variations may be derived. One operation by which this may be accomplished is contraposition. This operation consists in effecting a change in the order of the terms of a proposition, the state of things which the proposition is designed to express being supposed to remain unchanged. Thus the state of things expressed by the proposition "every A is a B" may be expressed also by "every non-B is a non-A," or by the form, "there is a B for every A."

We proceed now to apply this principle to the syllogism. For our notation let us take letters A, B, C, etc. for general terms, and express their negatives by writing dashes over them, $\bar{A}$, $\bar{B}$, $\bar{C}$, etc. Let a short curved mark over a letter indicate that its logical quantity has been changed; thus, $\breve{A}$, $\breve{B}$, $\breve{C}$, etc. A general term will be thus made particular, and a term already particular will be made general. Let us use the sign $-\!\!<$ for the copula.[1] We may then express the syllogism *Barbara* in the form

$$\begin{array}{l} A -\!\!< B \\ B -\!\!< C \\ \therefore A -\!\!< C \end{array}$$

[1] This notation is that used by Mr. C. S. Peirce, "On the Logic of Relatives." *Memoirs Am. Acad. of Arts and Sciences*, vol. ix, 1870.

From this as a starting-point we may produce formal variations by various modes of contraposition. The exhibition of two such forms will suffice.

(1) We may regard the logical quality of the terms and contrapose. The form $A \prec B$ then becomes $\bar{B} \prec \bar{A}$, or, "every non-B is a non-A."

(2) We may regard the logical quantity of the terms and contrapose. The form $A \prec B$ then becomes $\breve{B} \prec \breve{A}$. The latter form we may take to mean, "there is a B for every A," or "the B's include all the A's."

Applying these two kinds of contraposition to *Barbara*, we obtain the following variations:—

Fundamental Form.							
	Qualitative Variations.						
$A \prec B$	$\bar{B} \prec \bar{A}$	$A \prec B$	$\bar{B} \prec \bar{A}$	$A \prec B$	$\bar{B} \prec \bar{A}$	$A \prec B$	$\bar{B} \prec \bar{A}$
$B \prec C$	$B \prec C$	$\bar{C} \prec \bar{B}$	$\bar{C} \prec \bar{B}$	$B \prec C$	$B \prec C$	$\bar{C} \prec \bar{B}$	$\bar{C} \prec \bar{B}$
$\therefore A \prec C$	$\therefore A \prec C$	$\therefore A \prec C$	$\therefore A \prec C$	$\therefore \bar{C} \prec \bar{A}$	$\therefore \bar{C} \prec \bar{A}$	$\therefore \bar{C} \prec \bar{A}$	$\therefore \bar{C} \prec \bar{A}$
	Quantitative Variations.						
	$\breve{B} \prec \check{A}$	$A \prec B$	$\breve{B} \prec \check{A}$	$A \prec B$	$\breve{B} \prec \check{A}$	$A \prec B$	$\breve{B} \prec \check{A}$
	$B \prec C$	$\breve{C} \prec \breve{B}$	$\breve{C} \prec \breve{B}$	$B \prec C$	$B \prec C$	$\breve{C} \prec \breve{B}$	$\breve{C} \prec \breve{B}$
	$\therefore A \prec C$	$\therefore A \prec C$	$\therefore A \prec C$	$\therefore \breve{C} \prec \check{A}$	$\therefore \breve{C} \prec \check{A}$	$\therefore \breve{C} \prec \check{A}$	$\therefore \breve{C} \prec \check{A}$

These may be classed as two figures according as the conclusion has the fundamental or contraposed form; or they may be classed as four figures according as one or other, or both, or neither premise has been contraposed; or as eight figures, if we regard merely the relative position of the terms. The number of such variations may be indefinitely increased by admitting other modes of contraposition, or by starting from other syllogistic forms. All these variations may be easily produced by a mechanical contrivance. In order to secure this I have constructed a machine (Fig. 1) which presents to view three flaps in which are inserted cards containing the premises and conclusion of the syllogism which is to undergo transformation. Each flap, on making a half-revolution, presents its proposition in a contraposed form. The flaps terminate on one side of

FIG. 1.

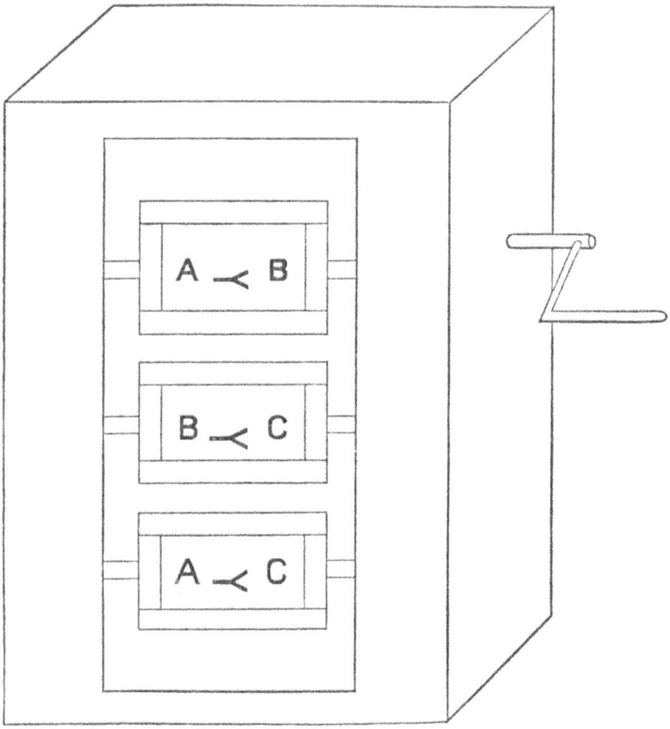

Scale ¼ in.

the machine in one-inch brass friction wheels. These are marked *a*, *b*, and *c* in Fig. 2. The wheels *d*, *e*, and *f* are, respectively, one, two, and four inches in diameter. Upon each of these wheels is fitted the sector of a wheel of like dimensions. Wheel *d* has on its outer side a sector of 180°; wheel *e*, on its inner side, one of 90°; wheel *f*, on its outer side, one of 45°. The friction of these sectors against the wheels *a*, *b*, and *c* causes the half-revolutions of the three flaps. By turning a crank attached to wheel *d*, the proposition A —< B is contraposed at the end of every turn, B —< C at every alternate turn, and A —< C at the end of every fourth turn. Eight turns of the crank will exhibit seven variations, and restore the fundamental syllogism to view.

FIG. 2.

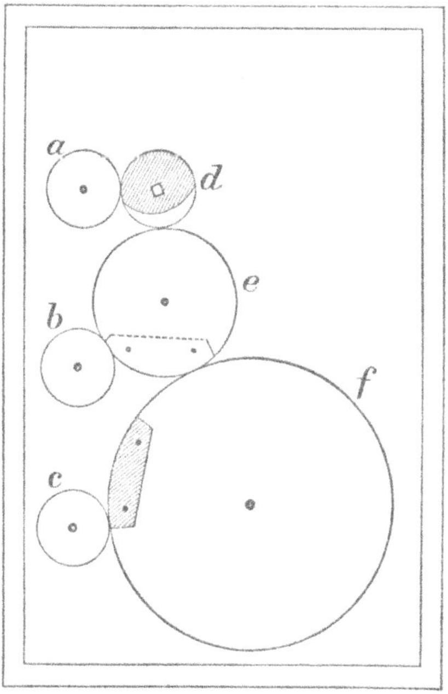

Scale ¼ in.

This mechanism could be readily extended so as to produce variations in a Sorites. A Sorites of *n* propositions would require, to contrapose its conclusion, a wheel of 2^{n-1} inches in diameter. We should secure, as in the syllogism, $2^n - 1$ variations for each kind of contraposition.

NOTE.—The Syllogistic Variation Machine will unfold to view the combinations of three logical terms and their negatives; or if we take the letters B — C, A — U, D — T, we obtain the words

B	C	B	C	B	C	B	C
A	A	U	U	A	A	U	U
D	D	D	D	T	T	T	T

NOTE ON AN EIGHT-TERM LOGICAL MACHINE.

I HAVE completed the design of an 8-term Logical Machine, of which a 4-term model is now nearly finished. If the premises be reduced to the form of the combinations to be excluded, as suggested by Boole and carried out by Venn, the operation of excluding these combinations may be performed mechanically by this machine. I have followed Jevons in making use of keys, but require for the 8-term machine only eight positive and eight negative letter keys and two operation keys. The excluded combinations are exhibited by indicators, which fall in the squares of one of my logical diagrams (Phil. Mag. ON. '81) from the perpendicular to a horizontal position. The non-excluded combinations, which constitute the conclusion, are exhibited by the indicators which are left standing.

ON THE ALGEBRA OF LOGIC.

By Christine Ladd.

There are in existence five algebras of logic, — those of Boole, Jevons, Schröder, McColl, and Peirce, — of which the later ones are all modifications, more or less slight, of that of Boole. I propose to add one more to the number. It will bear more resemblance to that of Schröder than to any of the others; but it will differ from that in making use of a copula, and also in the form of expressing the conclusion.[1]

ON IDENTICAL PROPOSITIONS.

The propositions which logic considers are of two kinds, — those which affirm the identity of subject and predicate, and those which do not. Algebras of logic may be classified according to the way in which they express propositions that are not identities. Identical propositions have the same expression in all. Of the logical theorems which are identities, I shall give those which are essential to the subject, and for the most part without proof.

(1) The sign $=$ is the sign of equality. $a = b$, a equals b, means that in any logical expression a can

[1] The substance of this paper was read at a meeting of the Metaphysical Club of the Johns Hopkins University, held in January, 1881.

be substituted for *b*, or *b* for *a*, without change of value. It is equivalent to the two propositions, "there is no *a* which is not *b*," and, "there is no *b* which is not *a*."

(2) The negative of a term or a proposition or a symbol is indicated by a line drawn over it. $\bar{a}$ = what is not *a*.

(3′) $a \times b$ = what is both *a* and *b*. As a class, it is what is common to the classes *a* and *b*. As a quality, it is the combination of all the qualities of *a* with all the qualities of *b*. When relative terms (XXI)[1] are excluded from consideration, *ab* may be written for $a \times b$.

(3°) $a + b$ = what is either *a* or *b*. As a class, it takes in the whole of *a* together with the whole of *b*, what is common to both being counted once only. It has the quality of either *a* or *b*, and hence the quality of the entire class is the quality common to *a* and *b*. The only qualities possessed by every member of the class "lawyers and bankers" are the qualities which lawyers and bankers have in common.

When arithmetical multiplication and addition are to be considered at the same time, logical multiplication and addition may be indicated by enclosing + and × in circles. The addition of logic has small connection with the addition of mathematics, and the multiplication has no connection at all with the process whose name it has taken. The object in borrowing the words and the signs is to utilize the familiarity which one has already acquired with processes which obey somewhat similar laws. There would not be the slightest difficulty in inverting the operations, and expressing logical multiplication in terms of addition, and logical addition in terms of multiplication. The essential processes of symbolic logic are either addition or multiplication (for greater convenience, both are used), and negation. The

[1] References in Roman numerals are to the titles at the end.

latter process renders any inverse processes which might correspond to subtraction and division quite unnecessary, and it is only on account of a supposed resemblance between the logical and the mathematical processes that an attempt to introduce them has been made.

(4′) $aaa = a.$	(4°) $a + a + \ldots = a.$
(5′) $abc = bca = cba.$	(5°) $a+b+c=b+c+a=c+b+a.$
(6′) $a(b+c) = ab + ac.$	(6°) $a + bc = (a+b)(a+c).$

The symbol ∞ represents the universe of discourse. (Wundt, Peirce.) It may be the universe of conceivable things, or of actual things, or any limited portion of either. It may include non-Euclidian n-dimensional space, or it may be limited to the surface of the earth, or to the field of a microscope. It may exclude things and be restricted to qualities, or it may be made coextensive with fictions of any kind. In any proposition of formal logic, ∞ represents what is logically possible; in a material proposition it represents what exists. (Peirce.) The symbol 0 is the negative of the symbol ∞; it denotes either what is logically impossible, or what is non-existent in an actual universe of any degree of limitation.

(7′) $a\bar{a} = 0.$	(7°) $a + \bar{a} = \infty.$
(8′) $a = a\,\infty = a(b+\bar{b})(c+\bar{c})\ldots$	(8°) $a = a + 0 = a + b\bar{b} + c\bar{c} + \ldots$
(9′) $\infty = a + \infty = a + (b+\bar{b}) + \ldots$	(9°) $0 = a\,0 = ab\bar{b}c\bar{c}\ \ldots$
(10′) $ab + a\bar{b} + \bar{a}b + \bar{a}\bar{b}$ $= (a+\bar{a})(b+\bar{b}) = \infty.$	(10°) $(\bar{a}+\bar{b})(\bar{a}+b)(a+\bar{b})(a+b)$ $= a\bar{a} + b\bar{b} = 0.$

The first member of this equation is called the complete development of two terms. The complete development of n terms, $(a+\bar{a})(b+\bar{b})(c+\bar{c})\ldots$, consists of the sum of 2^n combinations of n terms each.

(11′) $a + ab + abc + \ldots = a$	(11°) $a(a+b)(a+b+c)\ldots = a.$

This is called by Schröder the law of absorption.

The only process which presents any difficulty in this calculus is the process of getting the negative of a complex expression; and that difficulty is very slight if the right method is selected. There are three different methods, of which the last is of most frequent use. The first proceeds from the consideration that $ab + a\bar{b} + \bar{a}b + \bar{a}\bar{b}$ is a complete universe (10′), and that what is not one portion of a universe must be some other portion, if it exists at all. It follows that

$$\overline{ab} = a\bar{b} + \bar{a}b + \bar{a}\bar{b},$$

$$(12)\qquad \overline{ab + a\bar{b}} = \bar{a}b + \bar{a}\bar{b},$$

$$\overline{ab + a\bar{b} + \bar{a}b} = \bar{a}\bar{b},$$

and the process is the same for the complete development of any number of terms. This is the only rule made use of by Boole and by Mr. Jevons for obtaining a negative. If certain combinations of ten terms are given as excluded, to get those which are not excluded it is necessary, by this method, to examine 1,024 combinations of ten terms each.

The second method is contained in the following formulæ:—

(13′) $\overline{ab} = \bar{a} + \bar{b}.$	(13°) $\overline{a + b} = \bar{a}\bar{b}.$
$\overline{a\bar{b}} = \bar{a} + b.$	$\overline{a + \bar{b}} = \bar{a}b.$

That is, the negative of a product is the sum of the negatives of the terms, and the negative of a sum is the product of the negatives of the terms.[1] For example,

[1] Professor Wundt (XVIII., p. 257, note) makes the singular mistake of supposing that because $x(y+z) = xy + xz$, the parentheses must be removed before performing any general operation upon an expression. The negative of a product of the form $(a + b + c)\,m$, he says, is not $\bar{a}\bar{b}\bar{c} + \bar{m}$, but $(\bar{a} + \bar{m})(\bar{b} + \bar{m})(\bar{c} + \bar{m})$; and in working his problems he actually expresses it in this way, performs the indicated multiplication, obtaining $\bar{a}\bar{b}\bar{c} + (\bar{a} + \bar{b} + \bar{c})\,\bar{m} + \bar{m}$, and then reduces this expression by the absorption law (11′) to $\bar{a}\bar{b}\bar{c} + \bar{m}$.

$$\overline{a + bc + def} = \bar{a}(\bar{b} + \bar{c})(\bar{d} + \bar{e} + \bar{f}).$$

This rule was first given by De Morgan ("On the Syllogism," No. III., 1858). It may be proved in the following way:—

by (12),

$$\begin{aligned}\overline{ab} &= \bar{a}b + \bar{a}\bar{b} + a\bar{b} + \bar{a}\bar{b} \\ &= \bar{a}(b + \bar{b}) + (a + \bar{a})\bar{b} \\ &= \bar{a} + \bar{b}.\end{aligned}$$

$$\begin{aligned}\overline{a + b} &= \overline{a(b + \bar{b}) + (a + \bar{a})b} \\ &= \overline{ab + a\bar{b} + \bar{a}b} \\ &= \bar{a}\bar{b},\end{aligned}$$

by (12).

It appears that with the use of the negative sign the sum and the product are not both essential to complete expression. A sum can be expressed as the negative of a product, or a product can be expressed as the negative of a sum. The dualism which has been pointed out by Schröder, and which he indicates by printing his theorems in parallel columns, is, then, not an essential quality of things, but merely an accident of language. We prefer to say "what is either black or blue," to saying "what is not at the same time both not black and not blue;" but one is as easy to express symbolically as the other. It would not be difficult to develop the whole subject in terms of multiplication alone, or of addition alone; but the gain in simplicity is not equal to the loss in naturalness.

The third method of obtaining the negative of an expression is by means of the following equation:—

(14) $$\overline{pab + qa\bar{b} + r\bar{a}b + s\bar{a}\bar{b}} = \bar{p}ab + \bar{q}a\bar{b} + \bar{r}\bar{a}b + \bar{s}\bar{a}\bar{b}.$$

That is, consider any number of the letters as the elements of a complete development (10′), and take the negative of their coefficients. The reason is the same as for (12),—the two expressions together make up a complete universe, since

$$pab + \bar{p}ab = ab, \text{ etc.}$$

It is necessary to observe that if any part of the development is wanting, its coefficient is 0, and the negative of its coefficient is ∞. For instance,

$$\overline{(p+q+r)\,xy+stx\bar{y}+uvw\bar{x}y}$$
$$=p\bar{q}\bar{r}xy+(\bar{s}+\bar{t})\,x\bar{y}+(\bar{u}+\bar{v}+\bar{w})\,\bar{x}y+\bar{x}\bar{y}.$$

The entire number of combinations excluded by the first member is $7.2^5+2^6+2^5$, and that included by the second member is $2^5+3.2^6+7.2^5+2^8$, and together they make up 1024. This rule is given by Schröder only (XIV., p. 19). It is much easier of application than (12) or (13), except when the given expression bears no resemblance to a complete development.

(15) An expression may be said to be in its simplest form when it is represented by the smallest possible number of letters. It does not follow that it is then in its least redundant form. For instance, in

$$a+b, =a+\bar{a}b, =a\bar{b}+b,$$

$a+b$ is simpler than either of the other expressions, but it is redundant. It is

$$a\,(b+\bar{b})+(a+\bar{a})\,b,$$

which contains the combination ab twice; while

$$a+\bar{a}b, =a\,(b+\bar{b})+\bar{a}b,$$

contains each combination once only. The reduction of an expression to its simplest form may usually be accomplished by inspection. Take, for example, the expression

$$a+bc+\bar{a}\bar{b}d+\bar{a}\bar{c}d.$$

We have

$$a+\bar{a}\,(\bar{b}+\bar{c})\,d=a+\overline{bc}d,$$

and

$$bc+\overline{bc}d=bc+d.$$

Hence the whole expression is

$$a+bc+d.$$

If the reduction is not evident, it may be facilitated by taking the negative of the expression, reducing it, and then restoring it to the positive form (XVI., vol. x. p. 18).

ON THE COPULA.

I shall adopt the convention by which particular propositions are taken as implying the existence of their subjects, and universal propositions as not implying the existence of their subjects. Mr. Jevons would infer that the two propositions

> The sea-serpent is not found in the water,
> The sea-serpent is not found out of the water,

are contradictory; but Mr. McColl, Mr. Venn, and Mr. Peirce would infer that the sea-serpent does not exist. With this convention, contradiction can never exist between universal propositions nor between particular propositions taken by themselves. A universal proposition can be contradicted only by a particular proposition, and a particular only by a universal. The above premises are inconsistent with

> The sea-serpent has (at least once) been found.

With this convention, hypothetical and categorical propositions receive the same formal treatment. If a, then b = all a is b = a implies b. (Peirce.)

Algebras of Logic may be divided into two classes, according as they assign the expression of the "quantity" of propositions to the copula or to the subject. Algebras of the latter class have been developed with one copula only, — the sign of equality; for an algebra of the former class two copulas are necessary,[1] — one universal

[1] Every algebra of logic requires two copulas, one to express propositions of non-existence, the other to express propositions of existence. This necessarily follows from Kant's discussion of the nature of the affirmation of existence in the "Critik der reinen Vernunft." — *C. S. Peirce.*

and one particular. The following are the propositional forms which have been used by the principal recent writers on the algebra of logic:[1]—

	Traditional.	Boole and Schröder.	Jevons and Grassmann.	Grassmann.	McColl.	Peirce.
Universal.	All a is b	$a = vb$	$a = ab$	$a + b = b$	$a : b$	$a \prec b$
	No a is b	$a = v\bar{b}$	$a = a\bar{b}$	$a + \bar{b} = \bar{b}$	$a : \bar{b}$	$a \prec \bar{b}$
Particular.	Some a is b	$va = vb$	$ca = cab$	$ca + b = b$	$a \div \bar{b}$	$a \overline{\prec} \bar{b}$
	Some a is not b	$va = v\bar{b}$	$ca = ca\bar{b}$	$ca + \bar{b} = \bar{b}$	$a \div b$	$a \overline{\prec} b$

v is a special symbol, used to denote an arbitrary, indefinite class. It is immediately eliminated from the universal propositions, which then appear in the forms $a\bar{b} = 0$, $ab = 0$, respectively. In particular propositions "v is not quite arbitrary, and therefore must not be eliminated" (III., p. 124). Jevons makes no distinction between an indefinite class symbol, c, and any other class symbol. With Mr. McColl, every letter denotes a statement. By $a : b$ is meant that the statement that any object is a implies the statement that it is also b; but this does not affect the working of the algebra. The negative copula, $\div$, is the denial of the affirmative copula, :, and $a \div b'$, or, as he also writes it, $(a : b')'$, is read "a does not imply non-b." Mr. Peirce's symbol for the same copula is a modification of $\leq$. $a \overline{\prec} b$ is the denial of $a \prec b$, and is read, "a is not wholly contained under b." a and b may be either terms or propositions. The copula $\prec$ has an advantage over : in that it expresses an unsymmetrical relation by an unsymmetrical

[1] Mr. Venn has collected some two dozen ways in which "a is b" has been put into logical form.

symbol. $a \prec b$ may be written $b \succ a$ and read, "b contains a."

This quantified copula ($\prec$ or :) is positive for universal propositions, and negative for particular propositions. Another kind of quantified copula is possible,—namely, one which is particular when positive, and universal when negative. Instead of writing

$$\left.\begin{matrix} A \prec B \\ A \text{ is-wholly } B \end{matrix}\right\} \quad \text{and} \quad \left\{\begin{matrix} A \overline{\prec} B \\ A \text{ is-not-wholly } B \\ \text{or } A \text{ is-partly-not } B, \end{matrix}\right.$$

we might write

$$\left.\begin{matrix} A \mathbin{\overline{\vee}} B \\ A \text{ is-wholly-not } B \end{matrix}\right\} \quad \text{and} \quad \left\{\begin{matrix} A \vee B \\ A \text{ is-partly } B, \end{matrix}\right.$$

and it will appear that this latter plan has certain advantages. It comes perhaps a little nearer to common use. The sense "wholly" is usually attached to both *is* and *is-not*, but somewhat more strongly to the latter than to the former. We say, for instance, "flowers are fragrant," meaning that flowers are nearly always fragrant; but "leaves are not blue" means that leaves are absolutely never blue. "Knives are sharp" would be taken as true; "knives are not blunt" would excite opposition in the mind of the hearer.

The sign $\overline{\vee}$ is a wedge, sign of exclusion. $A \mathbin{\overline{\vee}} B$ is to be read "A is-not B," or "A is excluded from B." The sign $\vee$ is an incompleted wedge, sign of incomplete exclusion. $A \vee B$ is to be read "A is in part B," or "A is not-wholly excluded from B." $\vee$ is made into $\overline{\vee}$ by the addition of the negative sign; what is not not wholly excluded from anything is wholly excluded from it. $A \mathbin{\overline{\vee}} B$ and $A \vee B$ are contradictory propositions; each simply denies the other.

The eight propositions of De Morgan are then,—

$A \bar{\vee} B$	A is-not B; no A is B.
$A \vee B$	A is in part B; some A is B.
$A \bar{\vee} \bar{B}$	A is-not not-B; all A is B.
$A \vee \bar{B}$	A is partly not-B; some A is not B.
$\bar{A} \bar{\vee} B$	What is not A is-not B; A includes all B.
$\bar{A} \vee B$	What is not A is in part B; A does not include all B.
$\bar{A} \bar{\vee} \bar{B}$	What is not A is-not not-B; there is nothing besides A and B.
$\bar{A} \vee \bar{B}$	What is not A is in part not-B; there is something besides A and B, —

where $\vee$ connects terms that exist, while $\bar{\vee}$ connects terms which may be non-existent. Only six of these propositions are distinct, since there is no difference of form between $A \bar{\vee} \bar{B}$ and $\bar{A} \bar{\vee} B$, nor between $A \vee \bar{B}$ and $\bar{A} \vee B$.

Propositions expressed with the copula : or $-\!\!<$ are called inclusions; propositions expressed with the copula $\bar{\vee}$ may be called exclusions. Exclusions with an even number of negative signs are positive (affirmative) propositions; those with an odd number are negative propositions (De Morgan, "Syllabus of a Proposed System of Logic," p. 22). But the distinction, as Professor Wundt and others have pointed out, is unimportant. The only division of propositions which is of consequence is the division into universal and particular. The copulas $\bar{\vee}$ and $\vee$ are intransitive copulas, — a kind of copula of which De Morgan proposed to investigate the characters ("Syllabus," p. 31). They are symmetrical copulas, and the propositions $A \bar{\vee} B$, $A \vee B$, may be read either forward or backward. It is from the fact that there is no formal difference between subject and predicate that the advantages of this algebra follow. There is, however, a slight difference in meaning between $A \bar{\vee} B$ and $B \bar{\vee} A$;

the subject of the proposition is more evidently the subject of discourse. The propositions, "no men are mortal," and "there are no mortal men," convey the same information; but the first offers it by way of information about men, and the second by way of a description of the universe. Information may be given about a predicate by the use of a different kind of copula; as in "no lack of hospitality *is found among* Baltimoreans."

An inclusion is changed into the equivalent exclusion by changing the sign of the predicate. When an exclusion is to be made into an inclusion, it is a matter of indifference which of its terms is regarded as predicate; every exclusion contains within itself two inclusions, of which each is the converse by contraposition of the other. That is to say,

$$A \overline{\vee} B = A \prec \bar{B} = B \prec \bar{A},$$

$$A \vee B = A \overline{\prec} \bar{B} = B \overline{\prec} \bar{A}.$$

With this copula, therefore, the consideration of the conversion of propositions is rendered unnecessary. So also is the consideration of the quantification of the predicate. With the copula $\prec$ the subject and predicate have unlike quantity, or, more exactly, the quantity of the subject is universal and that of the predicate is indeterminate; $\prec$ means either equal to or less than. But with the copula $\overline{\vee}$ the quantity of both subject and predicate is universal, and with its denial $\vee$ both subject and predicate are taken in part only.

The copula $\prec$ must be taken in an inverted sense according as subject and predicate are taken in extension or in intension; but the copula $\overline{\vee}$ possesses the same meaning, whatever interpretation one gives to the terms which it separates. The proposition *men are animals* means that all the individuals, man, are contained among the individuals, animal; but that the qualities which distinguish an animal are contained among the qualities which distinguish a man. The proposition *no stones are plants* means that the objects denoted and the qualities connoted by the term *stone* are inconsistent with the objects denoted and the qualities connoted by the term *plant*. It is to be remembered that every term is at once a sum of objects and a product of qualities. If the term a denotes the objects $a_1, a_2, a_3 \ldots$ and connotes the qualities $\alpha_1, \alpha_2, \alpha_3 \ldots$, then

$$a = (a_1 + a_2 + \ldots)\,\alpha_1\alpha_2\alpha_3 \ldots$$

and the full content of the proposition *no a is b* is

$$(a_1 + a_2 + \ldots)\,\alpha_1\alpha_2\alpha_3 \ldots \;\overline{\vee}\;(b_1 + b_2 + \ldots)\,\beta_1\beta_2 \ldots$$

But the full content of the proposition *all a is b* can be expressed only by the two statements

$$a_1 + a_2 + \ldots + a_i \prec b_1 + b_2 + b_3 + \ldots \text{ and } \beta_1\beta_2 \ldots \beta_j \prec \alpha_1\alpha_2\alpha_3 \ldots$$

where the i objects a are identical each with some one of the objects b and the j qualities β are identical each with some one of the qualities α.

If p denotes a premise and c a conclusion drawn from it, then

$$p \;\overline{\vee}\; \bar{c} \qquad (m)$$

states that the premise and the denial of the conclusion cannot go together; and

$$p \vee \bar{c} \qquad (n)$$

states that the premise is sometimes accompanied by the falsity of the conclusion. It is hardly necessary to mention that (m) is satisfied by either the truth of the conclusion or the falsity of the premise, and that (n) implies that both the premise and the negative of the conclusion must, at some time, be true.

The word inference (or consequence) implies proceeding in a definite direction in an argument, — either from the premise to the conclusion, or from the negative of the conclusion to the negative of the premise. The argument $p \,\overline{\vee}\, c$ may be called an inconsistency. It is a form of argument into which the idea of succession does not enter; it simply denies the possible co-existence of two propositions. An inconsistency between two propositions is equivalent to each of two equivalent consequences, and a consistency to each of two equivalent inconsequences; or

$$p \,\overline{\vee}\, c = p -\!\!< \bar{c} = c -\!\!< \bar{p},$$

$$p \vee c = p \,\overline{-\!\!<}\, c = c \,\overline{-\!\!<}\, p.$$

The copulas $\overline{\vee}$ and $\vee$ with the symbol ∞ give means for expressing the total non-existence and the partial existence of expressions of any degree of complexity. Propositions with the symbol 0 do not occur in this algebra.

$(16')$ $x \,\overline{\vee}\, \infty$ = "x does not, under any circumstances, exist."

A universal proposition does not imply the existence of its subject; therefore $x \,\overline{\vee}\, 0$ = "x (if there is any x) is not non-existent," — a proposition which is true whatever x may be.

(16^0) $x \vee \infty$ = "x is at least sometimes existent."

A particular proposition does imply the existence of its subject; therefore $x \vee 0$ = "x exists, and at the same time does not exist," — a proposition which is false whatever x may be.

Since the symbol 0 will not appear at all in propositions expressed with these copulas, it will not be necessary to write the symbol ∞. I shall therefore express "there is no x" simply by $x \,\overline{\vee}$.

(17′) $a \bar{\vee} b = ab \bar{\vee}$

$abc \bar{\vee} = a \bar{\vee} bc = ca \bar{\vee} b = \ldots$

To say that no a is b is the same thing as to say that the combination ab does not exist.

(17°) $a \vee b = ab \vee$

$abc \vee = a \vee bc = ca \vee b = \ldots$

To say that some a is b is the same thing as to say that the combination ab does at least sometimes exist.

The factors of a combination which is excluded or not excluded may be written in any order, and the copula may be inserted at any point, or it may be written at either end. The proposition $abc \bar{\vee} de$ may be read "abc is-not de," "cd is-not abe," "abe is-not dc, — that is, is either not d or not c," etc. Any 0, 1, 2, 3, 4, or 5 of the letters may be made the subject and the others the predicate, and the positive or the negative universal copula may be used; or there are in all 2.32, = 64, different ways of putting the above proposition into words.

If a is a proposition, $a \bar{\vee}$ states that the proposition is not true in the universe of discourse. For several propositions, $abc \bar{\vee}$ means that they are not all at the same time true; and the way in which they are stated to be not all at the same time true depends on the character of the universe. If it be the universe of the logically possible, then $p_1 p_2 \bar{c} \bar{\vee}$ states that p_1 and p_2 may be taken as the premises and c the conclusion of a valid syllogism. It is the single expression in this system for a proposition which in the system of inclusions appears in the several forms

$$p_1p_2 -\!\!< c, \quad \bar{c} -\!\!< \bar{p}_1 + \bar{p}_2, \quad p_1 -\!\!< c + \bar{p}_2, \quad \bar{c}p_1 -\!\!< \bar{p}_2\,;$$

from the premises the conclusion follows; if the conclusion is false, one at least of the premises is false; from one premise may be inferred either the conclusion or the contradictory of the other premise, and from one premise and the contradictory of the conclusion follows the contradictory of the other premise. If the universe which is understood is the universe of what is possible in accordance with the laws of nature, then $ab \,\overline{\vee}$ denotes that the simultaneous truth of a and b is a contradiction of those laws. That x and y stand in the relation of cause and effect may be expressed by $x\bar{y} \,\overline{\vee}$. If x is a certain position and y its attendant acceleration, the above proposition states that the position and the absence of the acceleration are not found together; that from the position may be inferred the acceleration, and from the absence of the acceleration may be inferred the absence of the position. If $a \,\overline{\vee}\, \bar{b}$ means that Greeks are brave, and $c \,\overline{\vee}\, d$ means that the megatherium is not extinct, then

$$(a \,\overline{\vee}\, \bar{b}) \,\overline{\vee}\, (c \,\overline{\vee}\, d)$$

affirms that the co-existence of these two propositions is excluded from the universe of what is actually true. In like manner, according to the character of the universe of discourse, $a \vee b$ denotes either that the two propositions are logically consistent, or that they are possibly co-existent, or that they have actually been at some moment of time both true.[1]

ALGEBRA OF THE COPULA.

By the definition (1), we have

(18) $$(a = b) = (a \,\overline{\vee}\, \bar{b})\, (\bar{a} \,\overline{\vee}\, b).$$

Since also

$$(\bar{a} = \bar{b}) = (a \,\overline{\vee}\, \bar{b})\, (\bar{a} \,\overline{\vee}\, b),$$

it follows that

(19) $$(a = b) = (\bar{a} = \bar{b}).$$

In particular,

(20) $$(ab = 0) = (\overline{a\bar{b}} = \infty) = (ab \,\overline{\vee}\, \infty);$$

[1] The thorough-going extension of the idea of a limited universe to the relations between propositions is due to Mr. Peirce.

for the exclusions to which each equation is equivalent are

$$(ab \,\bar{\vee}\, \infty)\,(\overline{ab} \,\bar{\vee}\, 0),$$

and $\overline{ab} \,\bar{\vee}\, 0$ is a proposition of no content.

The principles of contradiction and excluded middle are therefore completely expressed by

(7′) $a\bar{a} \,\bar{\vee}$.

(7°) $\overline{a + \bar{a}} \,\bar{\vee}$.

In any symbolic logic there are three subjects for consideration,—the uniting and separating of propositions; the insertion or omission of terms, or immediate inference; and elimination with the least possible loss of content, or syllogism.

On uniting and separating Propositions.—From the definitions of logical sum and logical product applied to terms and to propositions we have the following identities:—

(21′) $(a \,\bar{\vee})(b \,\bar{\vee}) = (a + b \,\bar{\vee})$,

for the first member of the equation states that a does not exist and that b does not exist; and the second member states that neither a nor b exists.

(21°) $(a \vee) + (b \vee) = (a + b \vee)$,

for the first member of the equation states that either a exists or b exists; and the second member states that either a or b exists.

In both cases, a and b may be logical expressions of any degree of complexity. A combination of any number of universal propositions, or an alternation of any number of particular propositions, is then expressed as a single proposition by taking the sum of the elements of the separate propositions. This is the only form of inference (if it should be called inference at all) in which the conclusion is identical with the premises. The equations (21′) and (21°) are not in reality two distinct equations; they are, by (19), one and the same equation; since, by (13), the negative of $(a\,\overline{\vee})\,(b\,\overline{\vee})$ is $(a \vee) + (b \vee)$, and the negative of $a + b\,\overline{\vee}$ is $a + b \vee$. They are each equivalent to the two inconsistencies,

$$(21) \qquad \begin{array}{c} (a\,\overline{\vee})\,(b\,\overline{\vee})\,\overline{\vee}\,(a + b \vee) \\ (a \vee) + (b \vee)\,\overline{\vee}\,(a + b\,\overline{\vee}). \end{array}$$

There is no single expression in this algebra for a sum of universal propositions or a product of particular propositions.

To express that the propositions, some a is b and some c is d, are not both at the same time true (or that it is true throughout the universe of discourse that either no a is b or else no c is d), we must write

$$(a \vee b)\,(c \vee d)\,\overline{\vee},$$

To express that neither of the propositions, no a is b and no c is d, is true (or that it is true throughout the universe of discourse that both some a is b and some c is d), we must write

$$(a\,\overline{\vee}\,b) + (c\,\overline{\vee}\,d)\,\overline{\vee}.$$

And the expression for the corresponding particular propositions which follow from these universals is

$$(a\,\overline{\vee}\,b) + (c\,\overline{\vee}\,d) \vee;$$

that is, there is some time when either no a is b or else no c is d.

$$(a \vee b)\,(c \vee d) \vee;$$

that is, there is some time when both some a is b and some c is d.

On inserting and dropping Terms.—The following inconsistencies are immediate consequences of the definitions of the sum and the product:—

$$(22) \qquad (a + b + c\,\overline{\vee})\,\overline{\vee}\,(a + b \vee),$$

$$(23) \qquad (abc \vee)\,\overline{\vee}\,(ab\,\overline{\vee}).$$

The first asserts that the total non-existence of several things is inconsistent with the existence of some of them; the second asserts that the total non-existence of something, as *ab*, is inconsistent with the existence of some part of it, as *ab* which is *c*. They are not two distinct inconsistencies, however; either may be derived as a consequence from the other. These inconsistencies, when put into the form of inferences, become —

(22′) If $a + b + c \overline{\vee}$,
then $a + b \overline{\vee}$;

(23′) If $ab \overline{\vee}$,
then $abc \overline{\vee}$.

That is to say, given a universal exclusion, factors may be introduced and parts of a sum may be dropped, but not without loss of content.

(22°) If $a + b \vee$,
then $a + b + c \vee$;

(23°) If $abc \vee$,
then $ab \vee$.

That is to say, given a partial inclusion, factors may be dropped and parts of a sum may be introduced, but not without loss of content.

As a particular case of both of the inconsistencies (22) and (23) we have

$$(a \overline{\vee} b)(c \overline{\vee} d) \overline{\vee} (ac \vee b + d).[1] \qquad I.$$

If into the expression which is affirmed not to exist, $ab + cd$, we introduce the factor $c + a$; and if from the product, $acb + acd + ab + cd$, we drop the parts of a sum, $ab + cd$, — there remains $ac\,(b + d)$, the existence of which is inconsistent with the non-existence of *ab* and *cd*. Since there is no difference between subject and predicate,

$$(a \overline{\vee} b)(c \overline{\vee} d) \overline{\vee} (a + c \vee bd)$$

is an inconsistency of quite the same nature as *I.* For the expression of *I.* in words we have —

I_a. It is not possible that what is common to several classes should have any quality which is excluded from

[1] In its affirmative form, "if *a* is *b* and *c* is *d*, then *ac* is *bd*," this is Theorem I. of Mr. Peirce's paper on the Algebra of Logic (XXI.). As pointed out by Mr. Venn, it was first given by Leibnitz: "Specimen demonstrandi," Erdmann, p. 99.

one of them. If, for example, no bankers are poor and no lawyers are honest, it is impossible that lawyers who are bankers should be either poor or honest.

In this way the theorem is put into words in terms of a quality which is excluded from a class. It is a property of the negative copula that it lends itself equally well to the expression of propositions wholly in extension and wholly in intension, and also with the subject taken intensively and the predicate extensively. We should have in words, in these cases respectively—

I_b. If several classes are respectively excluded from several others, no part of what is common to them can be included in any of the others;

I_c. If several qualities are inconsistent respectively with several others, their combination is not consistent with any of the others.

I_d. It is not possible that a combination of several qualities should be found in any classes from each of which some one of those qualities is absent. If, for example, culture is never found in business men nor respectability among artists, then it is impossible that cultured respectability should be found among either business men or artists.

The inconsistency *I*. is the most general form of that mode of reasoning in which a conclusion is drawn from two premises, by throwing away part of the information which they convey and uniting in one proposition that part which it is desired to retain. It will be shown that it includes syllogism as a particular case. The essential character of the syllogism is that it effects the elimination of a middle term, and in this argument there is no middle term to be eliminated.

When combinations of any number of terms are given as excluded, a proposition with which they are inconsistent can be formed by taking any number of terms out of each and uniting them as a sum and denying their co-existence with the product of the terms which remain. If

$$ab\bar{c}\,\overline{\vee}, \qquad pl\bar{h}\,\overline{\vee},$$

affirm that no American bankers are uncharitable and that no Philadelphia lawyers are dishonest, then it is impossible that any Philadelphia bankers are either uncharitable Americans or dishonest lawyers; that any uncharitable and dishonest lawyers are either Philadelphians or American bankers; that any bankers who are also Philadelphia lawyers are either uncharitable Americans or dishonest, etc. Any, none, one, two, or three, terms from the first premise may be taken to form the sum with any, none, one, two, or three, terms from the second premise; there are, therefore, sixteen different conclusions to be drawn in this way from these two premises,—of which $ab\bar{c}ph\bar{l}\,\overline{\vee}$ is the least, since it has dropped the most information, and $ab\bar{c}+pl\bar{h}\,\overline{\vee}$ is the greatest, since it has dropped none of the information.

The inconsistency *I.* may be put into an inference in four different ways, according as both universals, one universal, one universal and the particular, or the particular alone, is taken as premise and the negative of what remains as conclusion. There are, therefore (when *I.* contains the smallest possible number of propositions), four distinct forms of inference, or progressive argument, with no middle term, in each of which the conclusion is a diminished conclusion. The factors of *I.* are, in general, one particular and any number of universals. The number of distinct forms of progressive argument which can be made out of an inconsistency between n propositions of which $n-1$ are universal, by taking 1, 2,.. or $n-1$ of the universal propositions with or without the particular proposition as premise and the negative of what remains as conclusion, is $2(n-1)$. Argument by way of inconsistencies, therefore, whatever may be thought of its naturalness, is at least $2(n-1)$ times more condensed than argument in the usual form.

When *I.* is made into an inference in such a way that one conclusion is drawn from two premises, we have,

if the premises are both universal,

$$(24') \qquad a \mathrel{\bar{\nabla}} b$$
$$c \mathrel{\bar{\nabla}} d$$
$$\therefore ac \mathrel{\bar{\nabla}} b + d$$

If no bankers have souls and no poets have bodies, then no banker-poets have either souls or bodies.

if the premises are one universal and one particular,

$$(24^\circ) \qquad a \mathrel{\bar{\nabla}} b$$
$$ac \vee b + d$$
$$\therefore c \vee d$$

If no Africans are brave and some African chiefs are either brave or deceitful, then some chiefs are deceitful.

On Elimination.—In (24′) there is no elimination, and in (24°) there is elimination of the whole of the first premise and part of the second. The most common object in reasoning is to eliminate a single term at a time,—namely, one which occurs in both premises. Each of these inferences gives rise to a form of argument, as a special case, by which that object is accomplished,—the premises being on the one hand both universal, and on the other hand one universal and the other particular. The inconsistency *I.* becomes, when d is equal to $\bar{b}$, and hence $b + d$ equal to ∞,

$$(a \mathrel{\bar{\nabla}} b)\,(c \mathrel{\bar{\nabla}} \bar{b})\,(ac \vee \infty) \mathrel{\bar{\nabla}},$$

or

$$(a \mathrel{\bar{\nabla}} b)\,(\bar{b} \mathrel{\bar{\nabla}} c)\,(c \vee a) \mathrel{\bar{\nabla}}. \qquad II.$$

Given any two of these propositions, the third proposition, with which it is inconsistent, is free from the term common to the two given propositions; a, b, and c are, of course, expressions of any degree of complexity. The propositions $ma \overline{\vee} x + y$, $\bar{x}\bar{y} \overline{\vee} c + n$, for instance, are inconsistent with $ma \vee c + n$; any number of terms may be eliminated at once by combining them in such a way that they shall make up a complete universe.

When any two of the inconsistent propositions in *II.* are taken as premises, the negative of the remaining one is the conclusion. There are, therefore, two distinct forms of inference with elimination of a middle term, special cases of (24′) and (24°). If we write x for the middle term, we have

(25′)

$$a \overline{\vee} x$$
$$b \overline{\vee} \bar{x}$$
$$\therefore\ ab \overline{\vee}.$$

The premises are

$$a (b + \bar{b})\, x \overline{\vee}$$
$$(a + \bar{a})\, b\bar{x} \overline{\vee};$$

and together they affirm that

$$ab\,(x + \bar{x}) + a\bar{b}x + \bar{a}b\bar{x} \overline{\vee},$$

or

$$ab + a\bar{b}x + \bar{a}b\bar{x} \overline{\vee}.$$

Dropping the information concerning x, there remains

$$ab \overline{\vee}.$$

The information given by the conclusion is thus exactly one half of the information given by the premises (Jevons).

(25°)

$$a \overline{\vee} x$$
$$b \vee x$$
$$\therefore\ b\bar{a} \vee.$$

The second premise is

$$bx\,(ax + \overline{ax}) \vee,$$

which becomes, since there is no ax,

$$bx\,(\bar{a} + \bar{x}) \vee,$$

or

$$bx\bar{a} \vee.$$

Dropping the information concerning x, there remains

$$b\bar{a} \vee.$$

This conclusion is equivalent to

$$b\bar{a} \vee x + \bar{x};$$

but the premises permit the conclusion

$$b\bar{a} \vee x;$$

hence the amount of information retained is exactly one half of the (particular) information given by the premises.

Elimination is therefore merely a particular case of dropping irrelevant information.

When a and b are single terms, (25′) is the doubly universal syllogism, and it is the single form in which that syllogism appears in this algebra. When it is translated into syllogism with an affirmative copula, it is necessary to consider the four variations of figure which are produced according as x or $\bar{x}$ is made subject or predicate. The eight moods in each figure correspond to the eight variations of sign which may be given to a, b, and x. All the rules for the validity of the doubly universal syllogism are contained in these: —

(1) The middle term must have unlike signs in the two premises.

(2) The other terms have the same sign in the conclusion as in the premises.

When a and b are single terms, (25^c) is the universal-particular syllogism, and it is the only form of that syllogism in this algebra. It can be translated into eight different forms of syllogism with unsymmetrical copula, according as x is made subject or predicate of either premise, and according as $\bar{a}$ or b is made subject of the conclusion. The eight moods of the major and minor particular syllogism in each figure correspond to the eight variations of sign which may be given to a, b, and x. All the rules for the validity of the universal-particular syllogism are contained in these: —

(1) The middle term must have the same sign in both premises.

(2) The other term of the universal premise only has its sign changed in the conclusion.

Those syllogisms in which a particular conclusion is drawn from two universal premises become illogical when the universal proposition is taken as not implying the existence of its terms.[1]

[1] McColl: Symbolical Reasoning, — *Mind*, no. xvii. Peirce: Algebra of Logic, — *Am. Journal of Math.*, vol. iii.

The argument of inconsistency,

$$(a \,\bar{\vee}\, b)\,(\bar{b} \,\bar{\vee}\, c)\,(c \vee a)\,\bar{\vee}, \qquad II.$$

is therefore the single form to which all the ninety-six valid syllogisms (both universal and particular) may be reduced. It is an affirmation of inconsistency between three propositions in three terms, — such that one of the propositions is particular, and the other two are universal; and such that the term common to the two universal propositions appears with unlike signs, and the other two terms appear with like signs. Any given syllogism is immediately reduced to this form by taking the contradictory of the conclusion, and by seeing that universal propositions are expressed with a negative copula and particular propositions with an affirmative copula. Thus the syllogisms Baroko and Bokardo,[1]

All *P* is *M*,	Some *M* is not *P*,
Some *S* is not *M*,	All *M* is *S*,
∴ Some *S* is not *P*,	∴ Some *S* is not *P*,

are equivalent respectively to the inconsistencies

$$(P \,\bar{\vee}\, \bar{M})\,(S \vee \bar{M})\,(S \,\bar{\vee}\, \bar{P})\,\bar{\vee},$$

$$(M \vee \bar{P})\,(M \,\bar{\vee}\, \bar{S})\,(S \,\bar{\vee}\, \bar{P})\,\bar{\vee}.$$

[1] If there were ever any occasion to use the mnemonic verses of syllogism, it might be worth while to put them into a form in which each word should bear the mark of its figure, as well as of its mood and its method of reduction. By some slight changes in the words, the first, second, third, and fourth figures might be indicated by the letters *r*, *t*, *l*, and *n* respectively: —

(*r*) Barbara, Cegare, Darii, Ferioque prioris.
(*t*) Cesate, Camestes, Festivo, Batoko secundæ.
(*l*) Tertia, Dalipi, Disalmis, Dalisi, Felapo.
(*l*) Bokalo, Feliso, habet; quarta insuper addit,
(*n*) Bamanip, Camenes, Dimanis, Fesanpo, Fesison.

It is then possible to give a perfectly general rule, easy to remember and easy of application, for testing the validity of any syllogism, universal or particular, which is given in words. It is this:—

Rule of Syllogism.—Take the contradictory of the conclusion, and see that universal propositions are expressed with a negative copula and particular propositions with an affirmative copula. If two of the propositions are universal and the other particular, and if that term only which is common to the two universal propositions has unlike signs, then, and only then, the syllogism is valid.

For instance, the syllogism—

Only Greeks are brave,
All Spartans are Greeks,
Therefore all Spartans are brave,

is equivalent to the inconsistency—

Non-Greeks are-not brave,
Spartans are-not non-Greeks,
Some Spartans are not-brave,

which fails to stand the test of validity in two respects,—the term brave appears with unlike signs and the term Greeks with like signs. The syllogism—

All men are mortal,
Some mortals are happy,
Therefore some men are happy,

is equivalent to the inconsistency—

Men are-not immortal,
Some mortals are happy,
Men are-not happy,

and it is not valid for the same reasons as before,—the

term mortal appears with unlike signs, and the term men with like signs.

When a, b, and x are expressions of any degree of complexity, (25′) and (25°) still furnish the only means for the elimination of x. For instance, if

$$(ab + \bar{c}d)\, x \,\overline{\vee}$$

and

$$(\bar{a} + c)\, \bar{x} + bf \,\overline{\vee},$$

then

$$(ab + \bar{c}d)\,(\bar{a} + c) + bf \,\overline{\vee},$$

or

$$abc + \bar{a}\bar{c}d + bf \,\overline{\vee},$$

is all that can be said without reference to x. And if

$$(ab + \bar{c}d)\, \bar{x} + bf \vee$$

and

$$(\bar{a} + c)\, x \,\overline{\vee},$$

then the conclusion, irrespective of x, is

$$(ab + \bar{c}d)\, \overline{\bar{a} + c} + bf \vee,$$

or

$$a\bar{c}\,(b + d) + bf \vee.$$

If the premises consist of propositions about propositions, then any proposition which it is desired to drop may be eliminated in accordance with these two rules.

Syllogisms are the inferences, with elimination, which are obtained by taking two of the propositions of *I.* as premises and the other as conclusion. When one proposition only is taken as premise, the conclusion is an alternation of propositions; and, as a special case, a single arbitrary term (instead of two or none) may be introduced. We have —

(26′) $a \mathbin{\overline{\vee}} b,$

$\therefore (ac \mathbin{\overline{\vee}} b + d) + (c \vee d);$

or, in words, if no a is b, then either no ac is either b or d, or else some c is d. If no Africans are brave, then either some chiefs are deceitful, or else no African chiefs are either brave or deceitful. When $c = x$, $d = \bar{b}$, this becomes

(27′) $a \mathbin{\overline{\vee}} b$

$\therefore (a \mathbin{\overline{\vee}} x) + (\bar{b} \vee x).$

If no Africans are brave, then either no Africans are Chinese or else some Chinese are not brave.

(26°) $ac \vee b + d;$

$\therefore (a \vee b) + (c \vee d);$

or, in words, if some ac is either b or d, then either some a is b or some c is d. If some African chiefs are either brave or deceitful, then either some Africans are brave or some chiefs are deceitful. When $b = \bar{d} = x$, this becomes

(27°) $ac \vee$

$\therefore (a \vee x) + (c \vee \bar{x}).$

If some lawyers are bankers, then either some lawyers are honest or some bankers are dishonest.

Inference from Universals to Particulars. — Diminished statement and that particular form of diminished statement which is syllogism are the only reasoning processes that are valid when a universe which contains nothing is included among possible universes, — that is, when it is taken as possible that both x and $\bar{x}$ may be at the same time non-existent. When that universe is excluded, — when the postulate "x and non-x cannot both be non-existent" is taken as true, — one other form of reasoning is possible. That postulate is expressed by

$$(x \mathbin{\overline{\vee}}) \mathbin{\overline{\vee}} (\bar{x} \mathbin{\overline{\vee}}), \qquad P.$$

which is equivalent to the two inferences, "if x does not exist, then non-x does exist," and "if non-x does not exist, then x does exist;" or, from the total non-existence of any expression whatever may be inferred the existence of some part at least of its negative. If

$a(b+c)\overline{\vee}$, then $\bar{a}+\bar{b}\bar{c}\vee$, and if $\bar{a}+\bar{b}\bar{c}\overline{\vee}$, then $a(b+c)\vee$; or,

$$(a(b+c)\overline{\vee})\overline{\vee}(\bar{a}+\bar{b}\bar{c}\overline{\vee}).$$

If x is a proposition, $a\vee b$, then non-x is its denial, $a\overline{\vee}b$; and the postulate states that a proposition cannot be both true and false at the same time.

From the proposition

$$ab\overline{\vee}$$

follows, in this way,

$$\overline{ab}\vee;\ \text{that is,}\ \bar{a}+\bar{b}\vee.$$

The complete convention in regard to the existence of terms is therefore: the particular proposition $a\vee b$ implies the existence of both a and b; the universal proposition $a\overline{\vee}b$ does not imply the existence of either a or b, but it does imply the existence of either $\bar{a}$ or $\bar{b}$. The necessity of the convention (if it should be called a convention) is even more evident when a and b are propositions; in that case it is equivalent to saying that two propositions cannot be true together unless each is at some time true, and that they cannot be not true together unless one or the other is at some time false.

Mr. McColl has pointed out that from "all a is b," "some a is b" does not follow, because there may not be any a. But from

$$a\bar{b}\overline{\vee}$$

it does follow that

$$\overline{a\bar{b}}\vee;\ \text{that is,}\ ab+\bar{a}b+\bar{a}\bar{b}\vee;$$

or from "all a is b" it does follow that one at least of the propositions "some a is b," "some not-a is b," "some not-a is not b," is true. From any universal proposition follows some one at least of the three particular propositions which it does not contradict. If a is known to exist, then "some a is b" follows from "all a is b" by a syllogism: —

$$a\bar{b}\ \bar{\vee}$$

$$aa \vee$$

$$\therefore\ ab \vee$$

From "no sea-serpents have gills" we cannot infer that there are some sea-serpents which are without gills, unless it is known that there are some sea-serpents; but we can infer that either there are some sea-serpents without gills, or there are some things, with or without gills, which are not sea-serpents, or else there is nothing in the universe.

RESOLUTION OF PROBLEMS.

Rule. — Express universal propositions with the negative copula and particular propositions with the affirmative copula, remembering that $a = b$ is equivalent to

$$a\bar{b} + \bar{a}b\ \bar{\vee},$$

and that its contradictory, a is not equal to b, is equivalent to

$$a\bar{b} + \bar{a}b \vee.$$

From a combination of universal propositions, the conclusion, irrespective of any term or set of terms to be eliminated, x, consists of the universal exclusion of the product of the coefficient of x by that of the negative of x, added to the excluded combinations which are free from x as given. If the premises include an alternation of particular propositions, the conclusion consists of the partial inclusion of the total coefficient of x in the particular propositions by the negative of that of x in the universal propositions, added to the included combinations which are free from x as given.

If there is any reason for expressing a universal conclusion with an affirmative copula or a particular conclusion with a negative copula, it can be done by taking any term or set of terms as subject and the negative of what remains as predicate.

The premises may also contain an alternation of any number of universal propositions. If either

$$(p \mathrel{\overline{\vee}} x) \text{ or } (q \mathrel{\overline{\vee}} x) \text{ or } (r \mathrel{\overline{\vee}} z),$$

and if at the same time

$$am \vee x,$$

then

$$am\,(\bar{p} + \bar{q} + \overline{rz}) \vee$$

is the conclusion irrespective of x. When a combination of particular propositions is included among the premises, the conclusion consists of a combination of the same number of particular propositions. From

$$(p \mathrel{\overline{\vee}} x)\ (q \mathrel{\overline{\vee}} x)$$

$$(a \vee x)\ (b \vee x),$$

may be inferred the two propositions,

$$(a \vee \bar{p}\bar{q})\ (b \vee \bar{p}\bar{q}).$$

From particular propositions by themselves no conclusion follows, otherwise than by simply dropping unnecessary information.

Particular premises may be attached to the universal premises by the conjunction *or* instead of the conjunction *and*. In that case no elimination is possible (except what can be done between the universal propositions by themselves), and a conclusion can be obtained only by means of the postulate, ***P***. If either ($a \mathrel{\overline{\vee}} b$ and $c \mathrel{\overline{\vee}} d$) or ($g \vee h$ and $i \vee j$), then the conclusions are $gh + \overline{ab} \vee$, $ij + \overline{ab} \vee$, $gh + \overline{cd} \vee$, $ij + \overline{cd} \vee$. In general, then, the premises may consist of a combination or an alternation of universal propositions (two cases), or of particular propositions (two cases), or a combination or an alternation of universal propositions united as a sum or a product to a combination or an alternation of particular propositions (eight cases).

It is apparent that logical notation would be improved by the addition of another sign, by means of which an alternation of universal and a combination of particular propositions might be expressed as a single proposition,— a sign such that

$$(p + x) \text{ sign } qy \text{ sign } rz \,\overline{\vee}$$

should mean that some one of the expressions $p+x$, qy, rz, is totally non-existent, and its contradictory,

$$(p + x) \text{ sign } qy \text{ sign } rz \vee,$$

should mean that all of these are, at least in part, existent.

The plan of treating a set of universal premises as a command to exclude certain combinations of the terms which enter them is due to Boole; no adequate extension of his method so as to take in particular propositions is possible, without the use of some device which shall be equivalent to a particular copula. Boole's method of elimination between universal propositions is to put x first equal to 0 and then to 1 in the given function, and to take the product of the results so obtained. The only difference between this rule and that which I have given (which is Prof. Schröder's) is that it first introduces x into those terms which are already free from it, and then proceeds to eliminate it from all. The value of the function

$$ax + b\bar{x} + c, \qquad \text{or} \qquad ax + b\bar{x} + c\,(x + \bar{x}),$$

for $x = 0$ (in this case $b + c$) is the coefficient of $\bar{x}$, and its value for $x=1$ (in this case $a+c$) is the coefficient of x. I have shown that the method is not an invention of modern times, but that it is nothing more than a rule for working the syllogism,

All b is x, No a is x, $\therefore$ No a is b,

when a, b, and x are not restricted to being simple terms. With the unsymmetrical copula, there are four different forms of pairs of universal propositions which make possible the elimination of x (XXI., p. 39), and for its elimination between a universal and a particular proposition it would be necessary to consider eight different forms, corresponding in all to the twelve distinct forms of syllogism.

If the result which remains after elimination is of the form

$$am + b\bar{m} + c\,\bar{\vee} \qquad (c)$$

(where m is the term in regard to which information is sought, and where all the letters are expressions of any degree of complexity), and if there is any reason for being dissatisfied with the conclusion as it stands,—"no m is a, no b is not m, and there is no c,"—m may be made subject and predicate respectively of two affirmative propositions, "all b is m, and all m is $\bar{a}$." If it be desired to express the conclusion without any repetition, then we must first state what is true without regard to m,—in this case,

$$ab + c\,\bar{\vee},$$

"there is no ab nor c,"—and then this information must be used to diminish the propositions in m. The identities

$$\begin{aligned} a &= a\,(ab + c + \overline{ab + c}) \\ b &= b\,(ab + c + \overline{ab + c}) \end{aligned} \qquad (8')$$

become, when there is no $a\bar{b} + c$,

$$a = a \, . \, \overline{ab + c} = a\bar{b}\bar{c},$$

$$b = b \, . \, \overline{ab + c} = b\bar{a}\bar{c};$$

and hence, instead of

$$a \,\overline{\vee}\, m, \qquad b \,\overline{\vee}\, \bar{m},$$

it is sufficient to write

$$a\bar{b}\bar{c} \,\overline{\vee}\, m, \qquad b\bar{a}\bar{c} \,\overline{\vee}\, \bar{m};$$

or, affirmatively,

All m is $b + c + \bar{a}$,

All $b\bar{a}\bar{c}$ is m.

Prof. Schröder expresses in terms of m such a conclusion as

$$am + b\bar{m} + c\,(m + \bar{m}) = 0,$$

by means of the formula

$$\begin{aligned} &[(a + c)\, m + (b + c)\, \bar{m} = 0] \\ &= [m = \text{all } (b + c) + \text{some } \overline{a + c}]\,[ab + c = 0]. \end{aligned} \qquad C.$$

The first factor of the second member of the equation is equivalent to the propositions,

All m is $b + c + \bar{a}\bar{c}$,

All $(b + c)$ is m,

Some $\overline{a + c}$ is m;

that is, it contains the propositions of the first member (the first diminished by $ab + c = 0$ and the second not), but it contains in addition the particular proposition "some $\overline{a + c}$ is m," which is a legitimate inference from "no $(a + c)$ is m" only if $\overline{a + c}$ is known to exist. A more condensed equational form of the conclusion $am + b\bar{m} + c \,\overline{\vee}$ is

$$(m = \text{all } b\bar{a}\bar{c} + \text{some } \bar{b}\bar{a}\bar{c})\;(ab + c = 0).$$

Boole reaches the same conclusion, (C), but he does

it by an extremely circuitous route. Nothing could well be simpler of application or more evident than this rule of Prof. Schröder's, and there is no reason why one should not place implicit confidence in it, in an algebra in which particular propositions are not taken as implying the existence of their terms. It contains the solution of what Mr. Jevons calls the "inverse logical problem," and which he solves by a process "which is always tentative, and consists in inventing laws and trying whether their results agree with those before us" (XXII., p. 252). It makes all reference to tables and machines quite superfluous. It seems to have been overlooked by the latest expositor of Boole's system,—Mr. Venn. He says that Boole's method of getting his conclusion is "a terribly long process; a sort of machine meant to be looked at and explained, rather than to be put in use;" and that if ever we do feel occasion to solve such a problem, it can be done most readily "by exercise, so to say, of our own observation and sagacity, instead of taking, and trusting to, a precise rule for the purpose of effecting it" (XXIII., p. 316).

But Boole's form for the conclusion (besides being not quite legitimate in this algebra) is not that which is most natural or most frequently useful. It is, moreover, suited only to a logic of extension, and it would be difficult to interpret intensively. The very simple device which may be substituted for it is to make use of the same method for getting back from excluded combinations to affirmative propositions which was employed in passing from the given affirmative propositions to the excluded combinations: if

$$\text{All } b \text{ is } m = b \,\overline{\vee}\, \bar{m},$$

then

$$b \,\overline{\vee}\, \bar{m} = \text{all } b \text{ is } m.$$

In this way the conclusions are given in the form which has been adopted by Mr. McColl. Complicated problems are solved with far more ease by Mr. McColl than by Mr. Jevons; but that is not because the method of excluded combinations is not, when properly treated, the easiest method. A method of implications, such as that of Mr. McColl, is without doubt more natural than the other when universal premises are given in the affirmative form, but the distinction which it preserves between subject and predicate introduces a rather greater degree of complexity into the rules for working it. An advantage of writing $abc \overline{\vee}$ instead of $abc = 0$ is that the copula can be inserted at any point in the excluded combination, and that elimination can be performed on the premises as they are given, when they have been expressed negatively, without first transposing all the members to one side. Without something corresponding to a contradictory copula, particular propositions cannot be treated adequately, and complicated propositions of either kind cannot be simply denied. With it, the contradictory of "all a is all b," that is, "it is not true that all a is all b," is $a\bar{b} + \bar{a}b \vee$; that is, "either some a is not b or some b is not a." And the contradictory of

$$abc + a\bar{b}c + \bar{a}\bar{b}c \overline{\vee}$$

is

$$abc + a\bar{b}c + \bar{a}\bar{b}c \vee;$$

that is, some one at least of the given combinations is in existence.

EXAMPLES.

1. (By Mr. Venn in *Mind* for October, 1876.) The members of a board were all of them either bondholders or share-holders, but no member was bond-holder

and share-holder at once; and the bond-holders, as it happened, were all on the board. What is the relation between bond-holders and share-holders?

Put

$$a = \text{member of board},$$
$$b = \text{bond-holder},$$
$$c = \text{share-holder}.$$

The premises are evidently

$$a \mathrel{\overline{\vee}} bc + \bar{b}\bar{c},$$
$$b \mathrel{\overline{\vee}} \bar{a};$$

and taking the product of the coefficient of a by that of $\bar{a}$, we have

$$b\,(bc + \bar{b}\bar{c}) \mathrel{\overline{\vee}},$$

or

$$bc \mathrel{\overline{\vee}}.$$

The required relation is, therefore,

No bond-holders are share-holders.

2. (XXII., p. 283.) What are the precise points of agreement and difference between two disputants, one of whom asserts that (1) space (a) = three-way spread (b), with points as elements (c) (Henrici); while his opponent holds that (2) space = three-way spread, and at the same time (3) space has points as elements?

$$(a = bc) = (a\bar{b} + a\bar{c} + \bar{a}bc \mathrel{\overline{\vee}}), \tag{1}$$
$$[a = b] = [a\bar{b} + \bar{a}b\,(c + \bar{c}) \mathrel{\overline{\vee}}], \tag{2}$$
$$a\bar{c} \mathrel{\overline{\vee}}. \tag{3}$$

They both assert that

$$a\bar{b} + a\bar{c} + \bar{a}bc \mathrel{\overline{\vee}},$$

and the second asserts in addition that

$$\bar{a}b\bar{c} \mathrel{\overline{\vee}};$$

that is, that a three-way spread which had not points as elements would be space.

3. (XVI., vol. x. p. 21.) From the premises

$$a\bar{x} \mathrel{\overline{\vee}} \bar{c}\,(\bar{d}+\bar{y})$$
$$bx \mathrel{\overline{\vee}} \bar{c}\,(\bar{d}+\bar{y})\,\bar{e}$$
$$\bar{a}\bar{b} \mathrel{\overline{\vee}} \bar{x}\,(\bar{d}+e)\,\bar{c}$$
$$a+b+c \mathrel{\overline{\vee}} \bar{x}\bar{y}$$

deduce a proposition containing neither x nor y.

The term y does not occur at all; hence $\bar{y}$ can be eliminated only by dropping the parts which contain it. There remain

$$a\bar{c}\bar{d} + \bar{a}\bar{b}\,(\bar{d}+e) \mathrel{\overline{\vee}} \bar{x},$$
$$b\bar{c}\bar{d}\bar{e} \mathrel{\overline{\vee}} x;$$

and taking the product of the first members we have

$$ab\bar{c}\bar{d}\bar{e} \mathrel{\overline{\vee}}.$$

4. (XXIII., p. 310.)

Given $\left.\begin{matrix} xy = a \\ yz = c \end{matrix}\right\}$, find xz in terms of a and c.

The equations are equivalent to the exclusions

$$xy\bar{a} + \bar{x}a + \bar{y}a \mathrel{\overline{\vee}},$$
$$yz\bar{c} + \bar{y}c + \bar{z}c \mathrel{\overline{\vee}};$$

and after elimination of y there remains

$$\bar{x}a + \bar{z}c + x\bar{a}c + za\bar{c} \mathrel{\overline{\vee}}. \qquad (p)$$

Collecting the predicates of xz and $\bar{x}\bar{z}$, we have

$$\left.\begin{aligned} xz &\mathrel{\overline{\vee}} \bar{a}c + a\bar{c}, \\ \bar{x}+\bar{z} &\mathrel{\overline{\vee}} ac. \end{aligned}\right. \qquad (q)$$

Prof. Schröder's formula, C, p. 49,

$$\text{If } m \mathrel{\overline{\vee}} x \text{ and } \bar{m} \mathrel{\overline{\vee}} y, \text{ then } m = \text{all } y + \text{some } \bar{x},$$

gives, in this case,

$$xz = \text{all } ac + \text{some } (ac + \bar{a}\bar{c})$$
$$= \text{all } ac + \text{some } \bar{a}\bar{c}.$$

If it were required to find $x\bar{z} + \bar{x}z$, we should have

$$x\bar{z} + \bar{x}z \mathrel{\overline{\vee}} ac,$$
$$xz + \bar{x}\bar{z} \mathrel{\overline{\vee}} \bar{a}c + a\bar{c};$$

whence

$$x\bar{z} + \bar{x}z = \text{all } (\bar{a}c + a\bar{c}) + \text{some } (\bar{a}c + \bar{c}\bar{a} + \bar{a}\bar{c})$$
$$= \text{all } (\bar{a}c + a\bar{c}) + \text{some } \bar{a}\bar{c}.$$

It is evident that (p) cannot be inferred from (q).

5. (*Educational Times*, Feb. 1, 1881, 6616. By W. B. Grove, B. A.) The members of a scientific society are divided into three sections, which are denoted by a, b, c. Every member must join one, at least, of these sections, subject to the following conditions: (1) Any one who is a member of a but not of b, of b but not of c, or of c but not of a, may deliver a lecture to the members if he has paid his subscription, but otherwise not; (2) one who is a member of a but not of c, of c but not of a, or of b but not of a, may exhibit an experiment to the members if he has paid his subscription, but otherwise not; but (3) every member must either deliver a lecture or perform an experiment annually before the other members. Find the least addition to these rules which will compel every member to pay his subscription or forfeit his membership, and explain the result.

Put $x =$ he must deliver a lecture, $y =$ he must perform an experiment, and $z =$ he has paid his subscription. Then the premises are

$$\bar{a}\bar{b}\bar{c} \,\overline{\vee} \qquad (a)$$

$$a\bar{b} + b\bar{c} + c\bar{a} \,\overline{\vee}\, x\bar{z} \qquad (1)$$

$$a\bar{c} + c\bar{a} + \bar{a}b \,\overline{\vee}\, y\bar{z} \qquad (2)$$

$$\bar{x}\bar{y} \,\overline{\vee}. \qquad (3)$$

It is required that $\bar{z}$ be excluded from all that part of the universe from which it has not already been excluded; namely, from the negative of

$$(a\bar{b} + b\bar{c} + c\bar{a})\, x + (a\bar{c} + c\bar{a} + \bar{a}b)\, y + \bar{a}\bar{b}\bar{c} + \bar{x}\bar{y},$$

which is, by the second rule for getting the negative,

$$(\bar{a}\bar{b}\bar{c} + abc + \bar{x})\,(\bar{a}\bar{b}\bar{c} + ac + \bar{y})\,(a + b + c)\,(x + y),$$

or

$$abcx + ac\bar{x}y.$$

Hence the desired "least addition to the rules" is

$$abcx + ac\bar{x}y \,\overline{\vee}\, \bar{z},$$

or, "No one who has not paid his subscription can be a member of all three sections and deliver a lecture, or of a and c and perform an experiment without lecturing."

6. (III., p. 237. Proposed for simpler solution by Mr. Grove, *Educational Times*, April 1, 1881.) A number of pieces of cloth striped with different colors were submitted to inspection, and the two following observations were made upon them: —

(a) Every piece striped with white (w) and green (g) was also striped with black (b) and yellow (y), and *vice versa*.

(*b*) Every piece striped with red (d) and orange (r) was also striped with blue (u) and yellow, and *vice versa.*

It is required to eliminate yellow, and to express the conclusion in terms of green.

The premises are

$$wg = by, \qquad dr = uy;$$

and by (18′) they are equivalent to the exclusions

$$wg\,(\bar{b} + \bar{y}) + by\overline{wg}\,\bar{\vee},$$
$$dr\,(\bar{u} + \bar{y}) + uy\overline{dr}\,\bar{\vee}.$$

Collecting the coefficients of y and $\bar{y}$ we have

$$b\overline{wg} + u\overline{dr}\,\bar{\vee}\,y,$$
$$wg + dr\,\bar{\vee}\,\bar{y};$$

and taking the product of the left-hand members we have

$$uwg\overline{dr} + bdr\,(\bar{w} + \bar{g})\,\bar{\vee},$$

which is to be added to that part of the premises which does not contain y; that is, to

$$w\bar{b}g + dr\bar{u}\,\bar{\vee}.$$

Concerning g we have

$$g\,\bar{\vee}\,w\,(\bar{b} + u\overline{dr}), \qquad bdr\,\bar{\vee}\,\bar{g};$$

or, with the affirmative copula, by (30),

$$g \prec \bar{w} + b\bar{u} + bdr, \qquad bdr \prec g.$$

The first is equivalent to Boole's conclusion when that is reduced by $dr\bar{u} = 0$. For the second Boole gives only $bdrwu \prec g$.

To solve this problem by Mr. Jevons's method, it would be necessary to write out the one hundred and twenty-eight possible combinations of seven terms, and to examine them all in connection with each of the premises. As Mr. Jevons himself says: "It is hardly possible to apply this process to problems of more than six terms, owing to the large number of combinations which would require examination" (XIII., p. 96).

7. (III., p. 146). From the premises

$$\bar{x}\bar{z}\,(\bar{v} + wy + \bar{w}\bar{y})\ \overline{\vee}$$

$$\bar{v}\,xw\,(y\bar{z} + \bar{y}z)\ \overline{\vee}$$

$$x\,(v + y)\,(zw + \bar{z}\bar{w})\ \overline{\vee}$$

$$(\bar{x} + \bar{v}\bar{y})\,(z\bar{w} + \bar{z}w)\ \overline{\vee}$$

it is required, first, to eliminate v; second, to express the conclusion in terms of x; third, in terms of y; fourth, to eliminate x; fifth, to eliminate y.

The terms which involve v are

$$\bar{x}\bar{z} + xw\,(y\bar{z} + \bar{y}z) + \bar{y}\,(z\bar{w} + \bar{z}w)\ \overline{\vee}\ \bar{v},\quad x\,(zw + \bar{z}\bar{w})\ \overline{\vee}\ v;$$

whence, taking the product of the left-hand members, we have only

$$xz\bar{y}w\ \overline{\vee}, \tag{a}$$

which is to be added to that part of the premises which does not contain v,—namely, to

$$\bar{x}\bar{z}\,(wy + \bar{w}\bar{y}) + xy\,(zw + \bar{z}\bar{w}) + \bar{x}\,(z\bar{w} + \bar{z}w)\ \overline{\vee}.$$

Collecting the parts which contain x and $\bar{x}$ we have

$$x\ \overline{\vee}\ zw + y\bar{z}\bar{w}, \tag{b}$$

$$\bar{x}\ \overline{\vee}\ z\bar{w} + \bar{z}w + \bar{z}\bar{w}\bar{y}. \tag{c}$$

The negative of the second member of (c) is, by (14), $zw + \bar{z}\bar{w}y$, hence, by (18′), these two exclusions are equivalent to the identity

$$x = z\bar{w} + \bar{z}w + \bar{z}\bar{w}\bar{y}, \tag{d}$$

or

$$\bar{x} = zw + y\bar{z}\bar{w}.$$

No part of the conclusion has been dropped in (*b*) and (*c*); hence the propositions which concern *y* may be taken from them. They are

$$y \,\overline{\vee}\, x\bar{z}\bar{w}, \qquad \bar{x}z\bar{w} \,\overline{\vee}\, \bar{y}, \qquad (e)$$

or

$$y \prec \bar{x} + z + w, \qquad \bar{x}\bar{z}\bar{w} \prec y.$$

These exclusions yield nothing upon the elimination of *y*; hence the only relation between *x*, *z*, and *w* is, from (*b*) and (*c*),

$$xzw + \bar{x}z\bar{w} + \bar{x}\bar{z}w \,\overline{\vee}. \qquad (f)$$

These conclusions are the same as those of Mr. McColl, and they are equivalent to those of Boole and Schröder. Prof. Wundt (XVIII., p. 356) accidentally omits (*a*) in getting the conclusions in regard to *y*, and they are in consequence altogether wrong. He remarks that Schröder has treated the problem in a partly coincident manner. I do not find that Mr. Jevons has treated it at all.

8. Six children, *a*, *b*, *c*, *d*, *e*, *f*, are required to obey the following rules: (1) on Monday and Tuesday no four can go out together; (2) on Thursday, Friday, and Saturday, no three can stay in together; (3) on Tuesday, Wednesday, and Saturday, if *b* and *c* are together, then *a*, *b*, *e*, and *f* must remain together; (4) on Monday and Saturday *b* cannot go out unless either *d*, or *c*, *e*, and *f* stay at home. *b* and *f* are first to decide what they will do, and *c* makes his decision before *a*, *d*, or *e*. Find (*α*) when *c* must go out, (*β*) when he must stay in, and (*γ*) when he may do as he pleases.

Let *a* be the statement that *a* goes out, and $\bar{a}$ the statement that he stays in, etc. Then we have for the first two premises

$$M + T \mathrel{\overline{\vee}} abcd + bcde + \ldots \qquad (1)$$

$$Th + F + S \mathrel{\overline{\vee}} \bar{a}\bar{b}\bar{c} + \bar{a}\bar{b}\bar{d} + \ldots \qquad (2)$$

The third premise excludes from certain days the combination in which b and c are both out or both in, except when a, b, e, and f are together; that is,

$$T + \overline{W} + S \mathrel{\overline{\vee}} (\bar{b}\bar{c} + bc)\,\overline{abef + \bar{a}\bar{b}\bar{e}\bar{f}}$$

$$\mathrel{\overline{\vee}} (\bar{b}\bar{c} + bc)\,(\bar{a} + \bar{b} + \bar{e} + \bar{f})\,(a + b + e + f)$$

$$\mathrel{\overline{\vee}} \bar{b}\bar{c}\,(a + b + e + f) + bc\,(\bar{a} + \bar{b} + \bar{e} + \bar{f});$$

or, finally,

$$T + \overline{W} + S \mathrel{\overline{\vee}} \bar{b}\bar{c}a + \bar{b}\bar{c}e + \bar{b}\bar{c}f + bc\bar{a} + bc\bar{e} + bc\bar{f}. \qquad (3)$$

The last premise is, for Monday,

$$M \mathrel{\overline{\vee}} bd\,(c + e + f). \qquad (4)$$

On Saturday, c, e, and f cannot all stay at home, by (2); therefore, this part of the premise is

$$S \mathrel{\overline{\vee}} bd. \qquad (4')$$

The first thing required is the elimination of a, d, and e. That part of the premises which is already free from those letters is

$$Th + F + S \mathrel{\overline{\vee}} \bar{b}\bar{c}\bar{f} \qquad (2')$$

$$T + \overline{W} + S \mathrel{\overline{\vee}} bc\bar{f} + \bar{b}\bar{c}f. \qquad (3')$$

Nothing can be eliminated between (1) and (2), because $MTh = 0$, etc.

For the same reason, d cannot be eliminated between (4) and (2); and therefore the premise (4) must be simply dropped. a and e can be eliminated at once by combining (3) with (1) and with (2). From (3) and (1), we have respectively

$$(T + \overline{W} + S)\,bc \mathrel{\overline{\vee}} \bar{a} + \bar{e},$$

$$(M + T)\,(bc + cd + \ldots) \mathrel{\overline{\vee}} ae;$$

and taking the product of the right-hand members and the sum of the left-hand members, we have

$$T(bc)\,\bar{\nabla}. \tag{5}$$

From (3) and (2) we have respectively

$$(T+\overline{W}+S)\,\bar{b}\bar{c}\;\bar{\nabla}\;a+e,$$

$$(Th+F+S)\,(\bar{b}+\bar{c}+\bar{d}+\bar{f})\;\bar{\nabla}\;\bar{a}\bar{e};$$

whence, in the same way,

$$S(\bar{b}\bar{c})\,\bar{\nabla}. \tag{6}$$

By combining (4′) with that part of (2) which does not contain $\bar{a}$, $\bar{e}$, or $\bar{b}$, and does contain d,—namely, with

$$Th+F+S\;\bar{\nabla}\;\bar{f}\bar{c}\bar{d},\text{—}$$

we obtain

$$S\;\bar{\nabla}\;b\bar{f}\bar{c}. \tag{7}$$

The conclusion required is then contained in (2′), (3′), (5), (6), and (7). But the information given in regard to S and T may be somewhat simplified by collecting their predicates. We have

$$S\;\bar{\nabla}\;\bar{b}\bar{c}\bar{f}+\bar{b}\bar{c}f+\bar{b}\bar{c}+bc\bar{f}+b\bar{c}\bar{f},$$

or

$$S\;\bar{\nabla}\;\bar{b}\bar{c}+b\bar{f}, \tag{8}$$

and

$$T\;\bar{\nabla}\;bc+\bar{b}\bar{c}f; \tag{9}$$

which with

$$Th+F\;\bar{\nabla}\;\bar{b}\bar{c}\bar{f}, \tag{2′′}$$

$$\overline{W}\;\bar{\nabla}\;bc\bar{f}+\bar{b}\bar{c}f, \tag{3′′}$$

form the entire conclusion. Collecting the subjects of c and $\bar{c}$, we have

$$(Th+F)\,\bar{b}\bar{f}+(T+W)\,\bar{b}f+S\bar{b}\;\bar{\nabla}\;\bar{c} \tag{a}$$

$$Tb+Wb\bar{f}\;\bar{\nabla}\;c \tag{b}$$

$$Sb\bar{f}\;\bar{\nabla}, \tag{c}$$

where the last proposition is already independent of *c*, and where *c* cannot be eliminated between (*a*) and (*b*). The conclusion may be expressed in words in this way: (*α*), if on Thursday or Friday *b* and *f* are both at home, or if on Tuesday or Wednesday *f* goes out without *b*, or if *b* stays at home on Saturday, then *c* must go out; (*β*), if *b* goes out on Tuesday, or if *b* goes out without *f* on Wednesday, then *c* must stay at home; (*γ*), whether *c* goes out or stays in, *b* does not go out without *f* on Saturday.

ON THE CONSTITUTION OF THE UNIVERSE.

The number of combinations in the complete development of *n* terms is 2^n. In any actual universe of things, any one of these combinations may be either present or absent; hence the number of different ways in which a universe may be made up out of *n* things is 2^{2^n}. The following Table gives the sixteen possible constitutions of the universe with respect to two terms. The sign 1 indicates the presence of the combination at the head of which it stands, 0 its absence. With the aid of the dual notation, applied to logical algebra by Mr. Franklin,[1] each case may be defined by a number; it is only necessary to attribute powers of two as weights to the different combinations, and to describe each arrangement by the sum of the weights of the combinations which are present in it. If we take the combinations of *a* and *b* in the order ab, $\bar{a}b$, $a\bar{b}$, $\bar{a}\bar{b}$, then 4, or 0100, denotes that the combination $a\bar{b}$ is present, and nothing else; 9, or 1001, that $\bar{a}\bar{b}$ and ab are present and $a\bar{b}$ and $\bar{a}b$ are absent, etc.

[1] Johns Hopkins University Circular, April, 1881.

$\bar{a}\bar{b}$ 8	$a\bar{b}$ 4	$\bar{a}b$ 2	ab 1	
0	0	0	0	0
0	0	0	1	1
0	0	1	0	2
0	0	1	1	3
0	1	0	0	4
0	1	0	1	5
0	1	1	0	6
0	1	1	1	7
1	0	0	0	8
1	0	0	1	9
1	0	1	0	10
1	0	1	1	11
1	1	0	0	12
1	1	0	1	13
1	1	1	0	14
1	1	1	1	15

If *a* is animal and *b* is black, then the 5th case is that of a universe made up of black animals and animals which are not black; in the 12th case the things which are wanting are black animals and black things which are not animals,—that is, there are no black things in this universe; the 15th case is the actual universe with respect to the terms animal and black; the 0-case is a universe in which nothing exists. If the material universe is the subject of discourse, and if a means matter and $\bar{b}$ means indestructible, then the existing state of things is described by 4; indestructible matter exists, and what is not indestructible matter does not exist. This Table is given by Jevons (XIII., p. 135); but he does not take account of non-existent terms, and hence all but seven of the sixteen cases (all but 6, 7, 9, 11, 13, 14, 15) are considered by him to be logical absurdities. If a and b are propositions, then case 9 is a universe in which they are true together and false together, and in which the time during which a is true is identical with the time during which b is true, either logically or extra-logically. The 0-case is a universe in which no proposition is true. Two cases the sum of whose characteristic numbers is 15,—as 5 and 10, or 0101 and 1010,—have been called by Prof. Clifford complementary cases: what exists in one is what does not exist in the other.

To exactly define the constitution of any universe, it is necessary to state, in regard to each combination, that it is present or that it is absent. The simple laws which every two terms obey are therefore four in number, being partly universal propositions and partly particular; except in the 0-case, where all the universal propositions are true, and in case 15, where all the particular propositions are true. The perfectly symmetrical universes are thus the universe in which there is nothing and that in which there is some of everything. For case 8, we have

$$(a \mathrel{\overline{\vee}} b)(\bar{a} \mathrel{\overline{\vee}} b)(a \mathrel{\overline{\vee}} \bar{b})(\bar{a} \vee \bar{b}),$$

and for case 13

$$(a \vee b)(\bar{a} \mathrel{\overline{\vee}} b)(a \vee \bar{b})(\bar{a} \vee \bar{b}).$$

When two simple or compound statements cannot be

converted into each other by any interchange between the terms which enter them (including negatives of terms), they are said to belong to different types. The universal propositions in two terms are of six different types. None, one, two, three, or four of them may be true, and it is only in the case where two are true that a difference of type is produced by the way in which the propositions are selected. Those two may be taken so that one letter has the same sign in both or not. Thus we may have either,

$$ab + a\bar{b} \overline{\vee},$$

that is,

$$a \overline{\vee};$$

or

$$ab + \bar{a}\bar{b} \overline{\vee},$$

that is,

$$a = \bar{b}.$$

The following Table gives the six types, the propositions which define them, and the universes which belong to each type :—

Type.	Universal.	Particular.	Cases.
I.	. . . .	$(a \vee b)(\bar{a} \vee b)(a \vee \bar{b})(\bar{a} \vee \bar{b})$	15
II.	$a \overline{\vee} b$	$(\bar{a} \vee b)\ (a \vee \bar{b})\ (\bar{a} \vee \bar{b})$	8, 4, 2, 1
III.	$a \overline{\vee}$	$(\bar{a} \vee b)\ (\bar{a} \vee \bar{b})$	12, 3, 10, 5
IV.	$a = b$	$(a \vee b)\ (\bar{a} \vee \bar{b})$	6, 9
V.	$a + b \overline{\vee}$	$\bar{a} \vee \bar{b}$	7, 11, 13, 14
VI.	$a + \bar{a} + b + \bar{b} \overline{\vee}$	. . . .	1

I. and VI. are complementary types; and so are II. and V. The universes complementary to III. and IV. are of types III. and IV. respectively. Six is the number of types of a universe in two terms, when all the particular propositions which the universal propositions do not deny are known to be true. If one takes account of combinations of alternations and alternations of combinations of both particular and universal propositions, the number of types is largely increased.

A race of beings which always completely defined its universe would have the above four-fold statements for its forms of expression. The eight propositions which are used by the race which exists are not complete definitions of a universe, but they are symmetrical; each has an eight-fold degree of ambiguity. "No *a* is *b*" denies the existence of the combination *ab*, but it leaves it doubtful whether, of the remaining combinations, none, any one, any two, or all three exist. "Some *a* is *b*," which affirms the existence of the combination *ab*, restricts the universe to some one of the eight cases,—1, 3, 5, 7, 9, 11, 13, 15. If, however, propositions are taken in the other sense,—if positive (affirmative) propositions are taken as implying the existence of the subject and negative not,—then they do not include all possible states of things with symmetry. The negative universal and the positive particular propositions cover eight cases each, as before; but of the positive universal $a \overline{\vee} \bar{b}$ takes in the four cases 1, 3, 9, 11, and $\bar{a} \overline{\vee} b$ the six cases 1, 4, 5, 9, 12, 13 only, and their contradictories, the negative particular, have respectively a twelve-fold and a ten-fold degree of ambiguity.

On the other hand, a race of beings which had the greatest possible variety of expression would be able to speak with any degree of ambiguity at pleasure. It would have a distinct propositional form for restricting the universe to any one, one of any two, one of any three, etc., of the possible cases; or its entire number of propositions in two terms would be 2^{16} or $2^{16}-1$, according as one counts or does not count the case in which nothing is said. All the 65,536 or 65,535 things which can be said without using any other terms than theologians and scientists, for instance, the existing race is able to say, without very much difficulty, by combinations and alternations of its Aristotelian and Morganic propositions. To say that either no scientists are theologians (0, 2, 4, 6, 8, 10, 12, 14), or some theologians are not scientists (3, 7, 11, 15), or some of those who are not theologians are scientists and some are not scientists (13), or else everybody is a theologian (1), is to make a statement of fourteen-fold ambiguity,—to limit the constitution of the universe under consideration to some case exclusive of 5 and 9. The contradictory of a statement of the form

$$(a \overline{\vee} b) + (\bar{a} \vee b) + (a \vee \bar{b})(\bar{a} \vee \bar{b}) + (\bar{b} \overline{\vee})$$

is, by (13),

$$(a \vee b)(\bar{a} \overline{\vee} b)(a \overline{\vee} \bar{b} + \bar{a} \overline{\vee} \bar{b})(\bar{b} \vee);$$

and to affirm that there are some theologians who are scientists, and that there are no theologians who are not scientists, and that either all scientists or else all non-scientists are theologians, and that not everybody is a theologian, is to affirm that either 5 or 9 furnishes the complete description of the universe with respect to the terms scientist and theologian.

In three terms the number of combinations is 2^3, the number of possible universes is $2^{2^3}, = 256$, and the number of possible propositions with all degrees of ambiguity is 2^{256}. The types of universal propositions have been given by Mr. Jevons (XIII., p. 140), but the number is increased when single terms as well as combinations are permitted to be non-existent. Prof. Clifford's method for obtaining types ("Essays and Lectures.— On the Types of Compound Statement involving Four Classes") is not difficult when applied to these terms. It takes account of terms which do not exist, and the number of types which he gives for four terms, 396, would be different on any other hypothesis. The problem would certainly be extremely difficult if such statements as Mr. Jevons calls contradictory were excluded. Prof. Clifford's solution takes account of combinations only of universal propositions. The number of types of alternations only, and of alternations and combinations of particular propositions only, is also 396, and the entire number is in this way raised to 4,396; but the determination of the number for mixed universal and particular propositions and for mixed alternations and combinations of them is still in the region of unsolved problems.

In three terms, the number of types of combinations of universal propositions is twenty-six,—six four-fold, eight less than four-fold, and eight more than four-fold. The types of more than four-fold statement may be obtained by taking those combinations which are not excluded by the types of less than four-fold statement.

LESS THAN FOUR-FOLD.

I.	0
II.	abc
III.	$abc + abc$
IV.	$abc + \bar{a}\bar{b}\bar{c}$
V.	$abc + \bar{a}\bar{b}\bar{c}$
VI.	$abc + a\bar{b}c + \bar{a}bc$
VII.	$abc + ab\bar{c} + \bar{a}\bar{b}c$
VIII.	$abc + a\bar{b}\bar{c} + \bar{a}b\bar{c}$

FOUR-FOLD.

IX.	$abc + \bar{a}bc + a\bar{b}c + ab\bar{c}$
X.	$abc + a\bar{b}c + ab\bar{c} + a\bar{b}\bar{c}$
XI.	$abc + ab\bar{c} + a\bar{b}c + \bar{a}b\bar{c}$
XII.	$abc + ab\bar{c} + \bar{a}\bar{b}c + \bar{a}\bar{b}\bar{c}$
XIII.	$abc + a\bar{b}\bar{c} + \bar{a}b\bar{c} + \bar{a}\bar{b}c$
XIV.	$abc + a\bar{b}\bar{c} + \bar{a}b\bar{c} + \bar{a}\bar{b}\bar{c}$

When condensed, these exclusions appear in the following form. The Arabic numbers give the corresponding types in Mr. Jevons's Table.

I.	. .	0	XXII.	. .	1
II.	8	abc	XXI.	. .	$a + b + c$
III.	2	ab	XX.	. .	$a + b$
IV.	12	$(ab + \bar{a}\bar{b})\,c$	XIX.	. .	$a\bar{b} + \bar{a}b + c$
V.	11	$abc + \bar{a}\bar{b}\bar{c}$	XVIII.	3	$a\bar{b} + b\bar{c} + c\bar{a}$
VI.	7	$(a + b)\,c$	XVII.	. .	$ab + c$
VII.	9	$ab + \bar{a}\bar{b}c$	XVI.	4	$a\bar{b} + \bar{a}b + abc$
VIII.	13	$abc + (a\bar{b} + \bar{a}b)\bar{c}$	XV.	15	$(a + b)c + (a\bar{b} + \bar{a}b)\bar{c}$

IX.	10	$ab + bc + ca$
X.	. .	a
XI.	5	$ac + b\bar{c}$
XII.	1	$a\bar{b} + \bar{a}b$
XIII.	14	$a\,(bc + \bar{b}\bar{c}) + \bar{a}\,(\bar{b}c + b\bar{c})$
XIV.	6	$abc + (\bar{a} + \bar{b})\,\bar{c}$

The exclusions

IV., XVIII., XI., XII., XIII., XIV.

are equivalent respectively to the identities

$ac = bc,\quad a = b = c,\quad ac = b\bar{c},\quad a = b,\quad a = bc + \bar{b}\bar{c},\quad ab = c.$

In these Tables, the letters may represent propositions as well as terms; of the 256 ways in which three propositions may be put together they give the 22 which are of distinct type. Case V., for instance, is the case in which three propositions, p_1, p_2, p_3, are affirmed to be not all at the same time true and not all at the same time false; or, in other words, it is known that some one of them is true and some one of them is false. In case XVII., p_1 and p_2 are not true together, and p_3 is not true at all. When the universe under discussion is the logical universe, the Tables serve to enumerate the 22 possible types of argument between three propositions. In case IX., p_1, p_2, p_3 are propositions so related that from the truth of any one the falsity of the other two can be inferred; in case XI., they are such that if two of them are both false or both true, the third is therefore false; and, conversely, if that is false, the others are therefore either both true or both false. The syllogism $p_1 p_2 p_3 \overline{\vee}$, is of the type II. The argument "if either some animals covered with fur are black or some black things not covered with fur are animals, then some animals are black," — that is,

$$(abx \vee) + (ab\bar{x} \vee)\ \overline{\vee}\,(ab\,\overline{\vee}),$$

which is of the form $(p_1 + p_2)\,\bar{p}_3\,\overline{\vee}$, — belongs to type VI.; and the identity,

$$(a\,\overline{\vee}\,b)\,(c\,\overline{\vee}\,d) = (ab + cd\,\overline{\vee}),$$

belongs to type XIV. In order to find actual arguments of all the 22 types, it would probably be necessary to go into some hyper-universe where the laws of thought are different from those under which we reason.

BIBLIOGRAPHY.

I. George Boole : The Mathematical Analysis of Logic, being an essay towards the Calculus of Deductive Reasoning. 1847.

II. Boole : The Calculus of Logic. *Cambridge and Dublin Math. Jour.*, Vol. III., 1848.

III. Boole : An Investigation of the Laws of Thought, on which are founded the Mathematical Theories of Logic and Probabilities. 1854.

IV. Hermann Grassmann : Lehrbuch der Arithmetik. Berlin, 1861.

V. W. S. Jevons : Pure Logic, or the Logic of Quality apart from Quantity. London and New York, 1864.

VI. C. S. Peirce : On an Improvement in Boole's Calculus of Logic. *Proc. Am. Acad. of Sciences*, Vol. VI., 1867.

VII. W. S. Jevons : The Substitution of Similars, the true Principle of Reasoning, derived from a modification of Aristotle's dictum. London, 1869.

VIII. C. S. Peirce : Description of a Notation for the Logic of Relatives. *Memoirs of the Am. Acad. of Sciences*, Vol. IX., 1870.

IX. Robert Grassmann : Die Formenlehre oder Mathematik. Zweites Buch : Die Begriffslehre oder Logik. Stettin, 1872.

X. A. J. Ellis : On the Algebraical Analogues of Logical Relations. *Proc. of the Royal Society of London*, 1872–73.

XI. Arthur Cayley : Note on the Calculus of Logic. *Quart. Jour. of Math.*, Vol. XI.

XII. Ernst Schröder : Lehrbuch der Arithmetik und Algebra für Lehrer und Studirende. I. Bd. : Die sieben algebraischen Operationen. Leipzig, 1873.

XIII. W. S. Jevons : Principles of Science. London, 1874. Third edition, 1879.

XIV. Ernst Schröder : Der Operationskreis des Logikkalkuls. Leipzig, 1877.

XV. J. Delboeuf : Logique Algorithmique. Liège et Bruxelles, 1877.

XVI. Hugh McColl : The Calculus of Equivalent Statements, and Integration Limits. *Proc. London Math. Soc.*, Vol. IX., 1877–78. Vol. X., 1878, Vol. XI., 1880.

XVII. Gottlob Frege : Begriffsschrift, eine der arithmetischen nachgebildete Formelsprache des reinen Denkens. Halle, 1879. Reviewed by Ernst Schröder in *Zeitschrift für Mathematik und Physik*, 1880.

XVIII. Wilhelm Wundt: Logik, eine Untersuchung der Principien der Erkenntniss und der Methoden wissenschaftlicher Forschung. I. Bd.: Erkenntnisslehre. Stuttgart, 1880.

XIX. J. Venn: On the Diagrammatic and Mechanical Representation of Propositions and Reasoning. *Phil. Mag.*, July, 1880.

XX. J. Venn: Symbolic Reasoning. *Mind*, July, 1880.

XXI. C. S. Peirce: On the Algebra of Logic. *Am. Jour. of Math.*, Vol. III., September, 1880.

XXII. W. S. Jevons: Studies in Deductive Logic. London, 1880.

XXIII. John Venn: Symbolic Logic. London, 1881.

NOTE.—In the foregoing article "combination" has been used as synonymous with "multiplication." In the following article, "combination" is used as including both multiplication and addition.

ON A NEW ALGEBRA OF LOGIC.

By O. H. Mitchell.

The algebra of logic which I wish to propose may be briefly characterized as follows: All propositions — categorical, hypothetical, or disjunctive — are expressed as logical polynomials, and the rule of inference from a set of premises is: *Take the logical product of the premises and erase the terms to be eliminated.* No set of terms can be eliminated whose erasure would destroy an aggregant term. So far as the ordinary universal premises are concerned, the method will be seen to be simply the negative of Boole's method as modified by Schröder. The reason is, that the terms which the propositions involve are virtually all on the right-hand side of the copula, instead of all on the left-hand side, as in Boole's method.

Attention is especially called to the treatment here given of particular propositions (of which there is introduced a variety of new kinds) which is homogeneous with that of universals, the process of elimination being precisely the same in each case. For the sake of clearness it may be well to state at the outset that I use addition in the modified Boolian sense, — that is, $x + y$ = all that is either x or y.

§ 1. Logic has principally to do with the relations of objects of thought. A *proposition* is a statement of such a relation. The objects of thought, among which relations may be conceived to exist, include not only class terms but also propositions. The statement of a relation among propositions is a proposition about propositions, which Boole called a secondary proposition. But every proposition in its ultimate analysis expresses a relation among class terms. The universe of class terms, implied by every proposition or set of propositions, may be limited or unlimited. Two class terms, a, b, are defined as the negatives of each other by the equations

$$a + b = U,$$
$$ab = 0,$$

where U is the symbol for the universe of class terms. Two propositional terms, α, β, are defined as the negatives or contradictories of each other by the equations

$$\alpha + \beta = \infty,$$
$$\alpha\beta = 0,$$

where ∞ is the symbol for the universe of relation, or for "the possible state of things." Mr. Peirce uses ∞ indifferently as a symbol for the universe of class terms, or for the universe of relation, but in the method of this paper it seems most convenient to have separate symbols. We can speak of "all of" or "some of" U, but hardly, it seems to me, of "all of" or "some of" the universe of relation; that is, the state of things. For this reason ∞ seems an especially appropriate symbol for the universe of relation.

The relation implied by a proposition may be conceived as concerning "all of" or "some of" the universe of class terms. In the first case the proposition is called universal; in the second, particular. The relation may be conceived as permanent or as temporary; that is, as lasting during the whole of a given quantity of time, limited or unlimited, — the Universe of Time, — or as lasting for only a (definite or indefinite) portion of it. A proposition may then be said to be universal or particular in time. The universe of relation is thus two-dimensional, so to speak; that is, a relation exists *among* the objects in the universe of class terms *during* the universe of time.

The ordinary propositions neglect the element of time; and these will first occupy our attention.

Let F be any logical polynomial involving class terms and their negatives, that is, any sum of products (aggregants) of such terms. Then the following are respectively the forms of the universal and the particular propositions: —

All U is F, here denoted by F_1,

Some U is F, " " F_u.

These two forms are so related that

$$F_1 + \bar{F}_u = \infty,$$

$$F_1\bar{F}_u = 0;$$

that is, F_1 and $\bar{F}_u$ are negatives of each other; that is, $\overline{(F_1)} = \bar{F}_u$. The two propositions F_1 and $\bar{F}_1$ satisfy the one equation

$$F_1\bar{F}_1 = 0,$$

and are "contraries" of each other. Whence, by taking the negative of both sides, we get

$$F_u + \bar{F}_u = \infty;$$

that is, F_u and $\bar{F}_u$ are "sub-contraries" of each other. The line over the F in the above does not indicate the negative of the proposition, only the negative of the predicate, F. The negative of the proposition F_1 is not $\bar{F}_1$, but $\overline{(F_1)}$, which, according to the above, $= \bar{F}_u$.

The Aristotelian propositions are represented in this notation as follows: —

$(\bar{a}+\bar{b})_1$	= All of U is $\bar{a}+\bar{b}$	= No a is b,	. . . .	E.
$(ab)_u$	= Some of U is ab	= Some a is b,	. . . .	I.
$(\bar{a}+b)_1$	= All of U is $\bar{a}+b$	= All a is b,	. . . .	A.
$(a\bar{b})_u$	= Some of U is $a\bar{b}$	= Some a is not b,	. .	O.

By substituting $\bar{a}$, $\bar{b}$ for a, b throughout we get the four complementary propositions of De Morgan. If these two forms be applied to the sixteen possible sums of ab, $a\bar{b}$, $\bar{a}b$, $\bar{a}\bar{b}$, there results the following

TABLE OF PROPOSITIONS.

$(ab + a\bar{b} + \bar{a}b + \bar{a}\bar{b})_1$	$(0)_u$
$(a\bar{b} + \bar{a}b + \bar{a}\bar{b})_1$	$(ab)_u$
$(\bar{a}b + \bar{a}\bar{b} + ab)_1$	$(a\bar{b})_u$
$(\bar{a}\bar{b} + ab + a\bar{b})_1$	$(\bar{a}b)_u$
$(ab + a\bar{b} + \bar{a}b)_1$	$(\bar{a}\bar{b})_u$
$(ab + a\bar{b})_1$	$(\bar{a}\bar{b} + \bar{a}b)_u$
$(ab + \bar{a}b)_1$	$(\bar{a}\bar{b} + a\bar{b})_u$
$(ab + \bar{a}\bar{b})_1$	$(\bar{a}b + a\bar{b})_u$
$(a\bar{b} + \bar{a}b)_1$	$(\bar{a}\bar{b} + ab)_u$
$(a\bar{b} + \bar{a}\bar{b})_1$	$(\bar{a}b + ab)_u$
$(\bar{a}b + \bar{a}\bar{b})_1$	$(a\bar{b} + ab)_u$
$(ab)_1$	$(a\bar{b} + \bar{a}b + \bar{a}\bar{b})_u$
$(a\bar{b})_1$	$(\bar{a}b + \bar{a}\bar{b} + ab)_u$
$(\bar{a}b)_1$	$(\bar{a}\bar{b} + ab + a\bar{b})_u$
$(\bar{a}\bar{b})_1$	$(ab + a\bar{b} + \bar{a}b)_u$
$(0)_1$	$(ab + a\bar{b} + \bar{a}b + \bar{a}\bar{b})_u$

Opposite propositions are negatives of each other. The Table reduced to its simplest form becomes

REDUCED TABLE.

1	$(U)_1$	$(0)_u$
2	$(\bar{a}+\bar{b})_1$	$(ab)_u$
3	$(\bar{a}+b)_1$	$(a\bar{b})_u$
4	$(a+\bar{b})_1$	$(\bar{a}b)_u$
5	$(a+b)_1$	$(\bar{a}\bar{b})_u$
6	$(a)_1$	$(\bar{a})_u$
7	$(b)_1$	$(\bar{b})_u$
8	$(ab+\bar{a}\bar{b})_1$	$(a\bar{b}+\bar{a}b)_u$
9	$(a\bar{b}+\bar{a}b)_1$	$(ab+\bar{a}\bar{b})_u$
10	$(\bar{b})_1$	$(b)_u$
11	$(\bar{a})_1$	$(a)_u$
12	$(ab)_1$	$(\bar{a}+\bar{b})_u$
13	$(a\bar{b})_1$	$(\bar{a}+b)_u$
14	$(\bar{a}b)_1$	$(a+\bar{b})_u$
15	$(\bar{a}\bar{b})_1$	$(a+b)_u$
16	$(0)_1$	$(U)_u$

If three terms be treated in a similar way we get $2.2^{2^3}, = 512$, different propositions. With n terms the total number is 2.2^{2^n}.

The propositions $(0)_1$ and $(0)_u$ assert that there is no universe of discourse, and are false in every argument, since a universe of class terms greater than zero is to be pre-supposed. Their negatives $(U)_u$, $(U)_1$ are therefore true in every argument. The eight propositions of De Morgan occur in lines 2, 3, 4, 5 of the Table.

Since the universe of class terms is supposed greater than zero, the *dictum de omni* gives

$$F_1 \overset{*}{-\!\!<} F_u;$$

that is, "all U is F" implies "some U is F."

To say "no U is F" is evidently the same as to say "all U is $\bar{F}$;" that is, $F_0 = \bar{F}_1$, and since a proposition whose suffix is 0 is thus expressible in a form with the suffix equal to 1, each suffix used will be supposed greater than zero. The suffix u in F_u is taken to be a *fraction* or *part* of U less than the whole; that is, "some of" U. In the proposition "some U is F" it is not denied that all U may be F, but the assertion is made of only a part of U. Thus u is taken as greater than zero and less than 1, or U. When u is written as a suffix of different propositions in the same argument, it is not meant that the *same* part of U is concerned in each case. F_1 is written for convenience instead of F_U. Sometimes F_ϵ will be written as a form inclusive of both the forms F_1 and F_u; that is, ϵ will be considered as having either of the two values 1 or u.

For *inference by combination* of such propositions we have the following simple rules:—

The conclusion from the *product* of two premises is the *product* of the predicates of the premises affected by a suffix equal to the *product* (in *extension*) of the suffices of the premises. Thus

$$F_\epsilon G_{\epsilon'} -\!\!< (FG)_{\epsilon\epsilon'}.$$

The conclusion from the *sum* of two premises is the *sum* of the predicates of the premises affected by a suffix equal to the *sum* (in *intension*) of the suffices of the premises. Thus

$$F_\epsilon + G_{\epsilon'} -\!\!< (F+G)_{\epsilon+\epsilon'}.$$

* This is Mr. Peirce's sign for the copula of inclusion, being an abbreviation of $\overline{\overline{<}}$. It is read "is," "is included under," or "implies." The following formulæ are sometimes made use of in this paper: $(a -\!\!< b) = (\bar{b} -\!\!< \bar{a}) = (a\bar{b} = 0) = (\infty = \bar{a} + b)$, where $\infty =$ the universe of discourse. Also, $(a -\!\!< b)\,(x -\!\!< y) -\!\!< (ax -\!\!< by)$.

When both premises are $\left\{\begin{matrix}\text{universal}\\ \text{particular}\end{matrix}\right\}$ the relation between the $\left\{\begin{matrix}\text{product}\\ \text{sum}\end{matrix}\right\}$ and the conclusion is equality; otherwise, the relation is $\prec$, an implication. Thus

(1) $F_1 G_1 = (FG)_1$,	$F_u + G_u = (F+G)_u$, (1′)
(2) $F_1 G_u \prec (FG)_u$,	$F_u + G_1 \prec (F+G)_u$, (2′)
(3) $F_u G_u \prec \infty$.	$F_1 + G_1 \prec (F+G)_1$. (3′)

These formulæ are so evident as hardly to need explanation. (1) means

$$(U = F)\,(U = G) = (U = FG),$$

and it follows from the definition of logical multiplication. By taking the negative of both sides, and changing $\bar{F}$, $\bar{G}$ to F, G, we get (1′). The law of the suffices in (1′) is $u + u = u$, or *some* + *some* = *some*. (2) means

$$(U = F)\,(u = G) = (u = FG),$$

and follows also from multiplication. The law of suffices is $1u = u$; that is, $Uu = u$. Since $G_1 \prec G_u$, (2′) follows from (1′). The law of the suffices is $u + 1 = u$; that is, "all of" *or* "some of" = "some of," which is addition in an intensive sense. In formula (3) there can be no inference when nothing is known about the relation of the two suffices; that is, $F_u G_u \prec \infty$. If it be known that u and u' have any common part, then for this common part $F_u G_{u'} \prec (FG)_{uu'}$. Thus if $u = \frac{3}{4}U$, and $u' = \frac{1}{2}U$, then $F_u G_{u'} \prec (FG)_{u''}$, where $u'' = uu' = \frac{1}{4}U$. Since we evidently have $(FG)_u \prec F_u G_u$, we get by contraposition the formula (3′), which means in words "'either all $U = F$, or all $U = G$' implies 'all $U =$ either F or G.'"

Having regard to (1) and (1′), it will be seen that

the most general proposition under the given conditions is of the form

$$\Pi\,(F_u + \Sigma G_1), \quad \text{or} \quad \Sigma\,(F_1 \Pi G_u),$$

where F and G are any logical polynomials of class terms, Π denotes a product, and Σ denotes a sum.

If F and G be any of the sixteen polynomials involving two class terms a, b, it is interesting to notice that any proposition, $\Sigma(F_1 \Pi G_u)$, can be reduced to the sum of products of the eight propositions of De Morgan. Thus, referring to the Table on page 76, any proposition F_1 in the first column is equivalent (1) to the product of one or more of the propositions 2, 3, 4, 5, — that is, E, A, E', A' (the two universal propositions added by De Morgan to the classic two being represented by E', A'); and any proposition G_u of the second column is equivalent (1′) to the sum of one or more of the propositions I, O, I', O', the two accented letters representing the particular propositions added by De Morgan. Thus $F_1 = \Pi\alpha$, and $\Pi G_u = \Pi\Sigma\beta = \Sigma\Pi\beta$, where α is one of the four universals of De Morgan, and β is one of the four particulars. Thus

$$\Sigma\,(F_1 \Pi G_u) = \Sigma\,(\Pi\alpha\Sigma\Pi\beta) = \Sigma\,(\Pi\alpha\Pi\beta).$$

Thus, for example, the proposition

$$(a+b)_u\,(ab+\bar{a}\bar{b})_1 + (a\bar{b})_1 + (a)_u\,(a+b)_1,$$

when reduced, becomes

$$\{(ab)_u + (a\bar{b})_u + (\bar{a}b)_u\}\,(\bar{a}+b)_1\,(a+\bar{b})_1$$
$$+\,(\bar{a}+\bar{b})_1\,(a+\bar{b})_1(a+b)_1 + \{(ab)_u + (a\bar{b})_u\}\,(a+b)_1;$$

i. e. $\quad AA'I + AA'O + AA'O' + EA'E' + E'I + E'O.$

In like manner it may be shown that if F, G, etc. be logical functions of any number of class terms, a, b, c, etc., the general proposition

$$\Pi\,(F_u + \Sigma G_1), \quad \text{or} \quad \Sigma\,(F_1 \Pi G_u)$$

may be reduced to a function of the eight propositions of De Morgan of the form

$$\Sigma\Pi\mu,$$

where μ, etc. are the eight propositions.

Propositions united by + form disjunctive propositions. A hypothetical proposition, "if α, then β," or $\alpha \prec \beta$, where α and β are themselves propositions, is evidently equivalent to the purely disjunctive proposition $\bar{\alpha} + \beta$. Thus "if a is bc, then $\bar{c}d$ is e" means

$$(\bar{a} + bc)_1 \prec (c + \bar{d} + e)_1;$$

which is the same as

$$(a\bar{b} + a\bar{c})_u + (c + \bar{d} + e)_1,$$

which may be put into words in one way as follows: "some a is either non-b or non-c, or all d which is non-c is e." The preceding formulæ are examples of *inference, by combination of propositions;* that is, of inference from a product or from a sum of propositions.

Inference by elimination will now be considered. It will only be necessary to consider the fundamental form F_ϵ, where ϵ may be either 1 or u. If F be a polynomial of the class terms, $a, b, c, \ldots x, y, z$, then x, y, z *may be eliminated from* F_ϵ *by erasure,* provided no aggregant term is thereby destroyed. That is,

$$F_\epsilon \prec F'_\epsilon,$$

where F' is what remains of F after the erasure. Thus

$$(\bar{a}x + bc\bar{x}y + d\bar{c}z + \bar{d}b)_\epsilon \prec (\bar{a} + bc + d\bar{c} + \bar{d}b)_\epsilon.$$

The reason is obvious. To say that "(all or some) U is $\bar{a}x$, or $bc\bar{x}y$, or etc.," is saying by an obvious implication that "(all or some) U is $\bar{a}$, or bc, or etc." F_ϵ means (all or some) $U \prec F$, and the erasure of a factor of a monomial term of F simply increases the extent of the term; therefore the predicate F is not diminished, and (all or some) $U \prec F'$, — that is, F_ϵ' is a valid inference. F' is really the sum of the coefficients of x, y, z in F, and is obviously a factor of F. The other factor of F is $F + \bar{F}'$; for $F'(F + \bar{F}') = F$, and $F + \bar{F}'$ is seen to contain no factor independent of x, y, z, since on erasing x, y, z, the result is $F' + \bar{F}'$, $= U$. If one of the aggregant terms of F contain no letters except those to be eliminated, then its coefficient is U, and F_ϵ' will in this case be a nugatory result. Thus from $(a + bcd)_\epsilon$ b, c, d, bc, bd, or cd can be eliminated, but not a, ab, ac, ad, abc, abd, acd, or bcd. As already stated, this algebra is the negative of Boole's as modified by Schröder, so far as *universal* premises are concerned. Thus Boole multiplied propositions by addition, and eliminated by multiplying coefficients. The method here employed multiplies propositions by multiplication, and eliminates by adding coefficients. When many eliminations are demanded in a problem, the advantage in point of brevity of this method over Boole's is of course greatly increased.

Before considering some illustrative examples, another kind of inference is to be noticed; namely, *inference by predication*; that is, the finding what a given proposition says about a given term, simple or complex. The rule is: *Multiply* F *by the given term*, m, *or add* $\bar{\text{m}}$ *to* F. *The resulting coefficient of* m *in* mF, *or the residue of* F *after adding* $\bar{\text{m}}$ *and reducing, will be the predicate of* m. Thus

$$F_\epsilon \prec (m = mF)_\epsilon,$$

$$\text{or} \quad F_\epsilon \prec (\bar{m} + F)_\epsilon.$$

The first means, "if $U = F$ for all or some U, then $m = mF$ for all or some U," and the result is obviously obtained by multiplying both sides of $U = F$ by m. The second relation means, "if $U = F$ for all or some U, then $U = \bar{m} + F$ for all or some U," and the result is obtained by adding $\bar{m}$ to both sides, remembering that $U + \bar{m} = U$. We have, of course,

$$(\bar{m} + F)_\epsilon = (\bar{m} + mF)_\epsilon = (m = mF)_\epsilon.$$

I now give the solution of the well-known problem of Boole, "Laws of Thought," p. 146. The premises are, remembering that $(a = b) = (\bar{a} + b)_1 \, (a + \bar{b})_1$,

$$(x + z + vy\bar{w} + vw\bar{y})_1,$$

$$(v + \bar{x} + \bar{w} + yz + \bar{y}\bar{z})_1,$$

$$(\bar{x} + \bar{v}\bar{y} + w\bar{z} + \bar{w}z)_1(xy + vx + wz + \bar{w}\bar{z})_1.$$

Multiplying the premises together, and dropping v from the result, we get

$$(wx\bar{z} + w\bar{x}z + \bar{w}xz + \bar{w}x\bar{y} + \bar{w}\bar{x}y\bar{z})_1, = \text{say } F_1.$$

The four results asked for by the problem are

(1) $$(\bar{x} + w\bar{z} + \bar{w}z + \bar{w}\bar{y})_1$$

(2) $$(w\bar{z} + wz + \bar{w}z + \bar{w}\bar{y} + \bar{w}y\bar{z})_1, \text{ i. e. } (U)_1$$

(3) $$(\bar{y} + \bar{w}\bar{x}\bar{z} + \bar{w}xz + w\bar{x}z + wx\bar{z})_1$$

(4) $$(\bar{w}x + \bar{w}\bar{z} + x\bar{z} + w\bar{x}z)_1.$$

The first gives the predicate of x in terms of y, z, w, being the same as $x \prec w\bar{z} + \bar{w}z + \bar{w}\bar{y}$, and is obtained by adding $\bar{x}$ to F and reducing. The second is the relation among y, z, w, and is obtained by dropping x from F and reducing. The result $(U)_1$ shows that no relation is implied among y, z, w alone. The third gives the predicate of y in terms of x, z, w, and is obtained by adding $\bar{y}$ to F and reducing. The fourth is the relation implied among x, z, w, and is obtained by dropping y from F and reducing. The relation (3) is not in its simplest form, since the implied relation (4) among x, z, w

has not yet been taken into account. Since (p. 81) we have $F = F'(F + \bar{F'})$, where F' is what remains of F after erasing y, and $F + \bar{F'}$ is that factor of F which contains no factor independent of y, we get $F_1 = F_1'(F + \bar{F'})_1$. The first factor F_1' is (4), and from the second factor we get $(\bar{y} + F + \bar{F'})_1$ as the simplest form of (3), that is,

$$(\bar{y} + w + \bar{x} + z.)_1.$$

Ordinary syllogism appears in this method as follows: The mood Barbara becomes

$$(\bar{a} + b)_1(\bar{b} + c)_1 = (\bar{a}\bar{b} + \bar{a}c + bc)_1 \prec (\bar{a} + c)_1,$$

b being eliminated by dropping it. The moods Darii, Datisi, Disamis, and Dimaris are all

$$(ab)_u\,(\bar{b} + c)_1 \prec (abc)_u \prec (ac)_u.$$

The premises of the mood Darapti are

$$(\bar{m} + p)_1(\bar{m} + s)_1, = (\bar{m} + sp)_1;$$

but there is no conclusion independent of the middle term m, since m cannot be eliminated. In inferring the conclusion I from these two premises logicians have virtually included a third premise $(m)_u$, that is, "some of U is m," or "there is some m." This with the product of the other two gives "some s is p;" that is,

$$(\bar{m} + sp)_1(m)_u \prec (spm)_u \prec (sp)_u.$$

In the same way, the premises of Felapton and Fesapo are

$$(\bar{m} + \bar{p})_1(\bar{m} + s)_1 = (\bar{m} + s\bar{p})_1,$$

and m cannot be eliminated here. With the additional premise $(m)_u$ we get "some s is not p;" that is,

$$(\bar{m} + s\bar{p})_1(m)_u \prec (s\bar{p}m)_u \prec (s\bar{p})_u.$$

The premises of Bramantip are

$$(\bar{p} + m)_1(\bar{m} + s)_1 = (s\bar{p} + sm + \bar{m}\bar{p})_1 \prec (s + \bar{p})_1;$$

that is, the conclusion is not "some s is p," but "all p is s," or "all $\bar{s}$ is $\bar{p}$," the proposition A'. Here, again, the conclusion "some s is p" has been reached only by the virtual inclusion of a third premise, "there is some p," that is, $(p)_u$. Then we have

$$(s\bar{p} + sm + \bar{m}\bar{p})_1 (p)_u \prec (smp)_u \prec (sp)_u .$$

This is the same thing as to say that a particular conclusion cannot be drawn from universal premises, since a particular proposition implies the existence of its subject, while a universal does not.[1]

As another illustration of the method, I solve the problem in Boole's "Laws of thought," p. 207. The premises are

$$(\bar{w} + x\bar{y}\bar{z} + \bar{x}y\bar{z} + \bar{x}\bar{y}z)_1 ,$$

$$(\bar{y} + p\bar{q}\bar{r} + \bar{p}q\bar{r} + \bar{p}\bar{q}r)_1 ,$$

$$(pq + \bar{p}\bar{q} + \bar{t})_1 ,$$

$$(\bar{x} + \bar{t})_1 ,$$

$$(\bar{t} + w)_1 .$$

Their product is

$$[\bar{t}\{\bar{w}\bar{y} + \bar{w}(p\bar{q}\bar{r} + \bar{p}q\bar{r} + \bar{p}\bar{q}r) + x\bar{y}\bar{z} + \bar{x}\bar{y}z + \bar{x}y\bar{z}(p\bar{q}\bar{r} + \bar{p}\bar{q}r + \bar{p}\bar{q}r)\} + w\bar{x}\{pq\bar{y}z + \bar{p}\bar{q}\bar{y}z + \bar{p}\bar{q}ry\bar{z}\}]_1, \text{ say } F_1 ,$$

which contains everything implied in the premises. The results asked for are

1. $(r + \bar{t} + z)_1$, whence $t \prec r + z$
2. $(r + \bar{t} + \bar{y})_1$, " $t \prec r + \bar{y}$
3. $(U)_1$,

[1] Mr. Peirce and others.

4.	$(\bar{t} + \bar{x})_1,$	whence	$t \prec \bar{x}$
5.	$(\bar{p} + \bar{q} + \bar{y})_1,$	"	$y \prec \bar{p} + \bar{q}$
6.	$(\bar{t} + \bar{y}z + y\bar{z}r)_1,$	whence	$\begin{cases} y\bar{z} \prec r + \bar{t} \\ \bar{r}t \prec \bar{y}z \\ t\bar{z} \prec yr \\ \bar{r}y \prec \bar{t} \end{cases}$
7.	$(\bar{t} + \bar{y}z + y\bar{z})_1,$	whence	$\begin{cases} \bar{t} \prec U \\ yz + \bar{y}\bar{z} \prec \bar{t}. \end{cases}$

The relations in the first column are each obtained by dropping from F_1 the letters not concerned in the quæsitum. Each predicate in the second column is obtained by multiplying its opposite F' by its subject. The result 4 disagrees with that obtained by Boole.

The two examples taken from Boole have dealt exclusively with universal propositions. The following is of a different kind: —

What may be inferred independent of x *and* y *from the two premises, " either some* a *that is* x *is not* y, *or all* d *is both* x *and* y *;" and " either some* y *is both* b *and* x, *or all* x *is either not* y *or* c *and not* b " *?*

The premises are

$$(ax\bar{y})_u + (\bar{d} + xy)_1,$$

$$(bxy)_u + (\bar{x} + \bar{y} + \bar{b}c)_1.$$

By multiplication we get

$$(ax\bar{y})_u\,(bxy)_u + (bxy)_u + (ax\bar{y})_u + (\bar{d}\bar{x} + \bar{d}\bar{y} + \bar{b}c\bar{d} + \bar{b}cxy)_1.$$

Whence, dropping x, y and reducing, we get

$$(b + a)_u + (\bar{d} + \bar{b}c)_1,$$

which may be interpreted in words, " there is some b, or a, or else all d is c and not b."

From this result we may further eliminate c. Eliminating c, we get

$$(b + a)_u + (\bar{d} + \bar{b})_1,$$

which means "either b or a exists, or no d is b."

The analogy between class and propositional terms.— Hitherto in the consideration of F_1 and F_u the polynomial F has been regarded as a function of class terms a, b, etc. Suppose a, b, etc. to be propositional terms like F_1 and F_u, and call the resulting polynomial no longer F, but Φ. Then the suffices of Φ_1 and Φ_u cannot be interpreted any longer as referring to the universe of class terms, since the propositional terms F_1, F_u, etc., of which Φ is a function, are supposed to have already suffices with this meaning. The suffices of Φ_1 and Φ_u can only be interpreted then as referring to the universe of the time during which the complex or secondary proposition Φ is supposed to be true. Then, if V denote the universe of time,

Φ_1 means	"Φ, during all	V,"	or	"all $V \prec \Phi$,"
Φ_v "	"Φ, "	some V,"	or	"some $V \prec \Phi$."

In other words

Φ_1 means	"Φ is always true,"
Φ_v "	"Φ is sometimes true,"

where "always" refers to the universe of time, V.

Owing to the similarity between class terms and propositional terms with respect to the operations of multiplication and addition, it follows that all that has been said in regard to inference from propositions like F_1, F_u holds equally with Φ_1 and Φ_v. Thus

$$\Phi_1\Psi_1 = (\Phi\Psi)_1, \qquad \Phi_v + \Psi_v = (\Phi + \Psi)_v,$$
$$\Phi_1\Psi_v \prec (\Phi\Psi)_v, \qquad \Phi_1 + \Psi_v \prec (\Phi + \Psi)_v,$$
$$\Phi_v\Psi_v \prec \infty, \qquad \Phi_1 + \Psi_1 \prec (\Phi + \Psi)_1.$$

So in regard to elimination, any set of terms can be eliminated by neglect, provided no aggregant term is thereby destroyed.

§ 2. *Propositions of Two Dimensions.*

Let U stand for the universe of class terms, as before, and let V represent the universe of time. Let F be a polynomial function of class terms, a, b, etc. Then let us consider the following system of six propositions:

F_{uv}, meaning "some part of U, during some part of V, is F,"
F_{u1}, " "some part of U, during every part of V, is F,"
F_{1v}, " "every part of U, during some part of V, is F,"
$F_{u'1}$, " "the same part of U, during every part of V, is F,"
$F_{1v'}$, " "every part of U, during the same part of V, is F,"
F_{11}, " "every part of U, during every part of V, is F."

By thus introducing the element of time, three varieties of the proposition F_u are distinguished,—F_{uv}, F_{u1}, $F_{u'1}$. Thus in speaking of the people of a certain village during a certain summer ($U =$ village, $V =$ summer), "some of the Browns were at the sea-shore during the summer" may mean either that some of them were there during a part of the summer, or that some of them were there during every part of the summer,—not necessarily the same persons,—or that the *same* persons were there during the whole summer. These three meanings are here denoted respectively by $(bs)_{uv}$, $(bs)_{u1}$, $(bs)_{u'1}$. Three varieties of F_1 are also distinguished,—F_{11}, F_{1v}, $F_{1v'}$. Thus "all the Browns were ill during the year" may mean either that every one was ill during every part of the year, or that every one was ill during some part of the year,—not necessarily the same part,—or that every one was ill during the *same* part of the year. These three meanings are denoted respectively by $(\bar{b} + i)_{11}$, $(\bar{b} + i)_{1v}$, $(\bar{b} + i)_{1v'}$.

The *dictum de omni* gives the following relations among these six propositions:—

$$F_{11} \prec F_{1v'} F_{u'1} F_{1v} F_{u1} F_{uv}, \text{ and } F_{11} + F_{1v'} + F_{u'1} + F_{1v} + F_{u1} \prec F_{uv};$$

and since *same* is included under *some*, we have

$$F_{1v'} \prec F_{1v}, \text{ and } F_{u'1} \prec F_{u1}.$$

The following pairs of propositions,

$$F_{uv} \text{ and } \bar{F}_{11}, F_{u1} \text{ and } \bar{F}^{*}_{1v'}, F_{u'1} \text{ and } \bar{F}_{1v},$$

satisfy the two equations

$$\alpha + \beta = \infty,$$
$$\alpha\beta = 0,$$

and the members of each pair are therefore the negatives or contradictories of each other. Thus if $F = \bar{b} + i$, it is seen that $(b\bar{i})_{uv}$ and $(\bar{b} + i)_{11}$ are contradictories; that is, "either some of the Browns were not ill during some part of the year, or they were all ill during every part of the year," and both cannot be true. An example of the second pair is $(bi)_{u1}$ and $(\bar{b} + \bar{i})_{1v'}$; that is, "either some of the Browns were ill during every part of the year (not necessarily the same persons during the whole year) or at some particular time none of them were ill," and both cannot be true. An example of the third pair is $(bi)_{u'1}$ and $(\bar{b} + \bar{i})_{1v}$, "either the same Browns were ill during the whole year, or it was true for each

* The natural first thought is that F_{11}, F_{u1}, F_{1v}, F_{uv} form a system of propositions by themselves, but it is seen that $F_{1v'}$ and $F_{u'1}$ must be added to the system, in order to contradict F_{u1} and F_{1v}. Mr. Peirce pointed out to me that these propositions are really *triple relatives*, and are therefore six in number. F_{11}, for instance, means "F is a description *of* U *during* V." See the Johns Hopkins University Circular, August, 1882, p. 204.

part of the village during some part of the year that none of the Browns were ill," and both cannot be true.

Since from $A \prec B$ we get $\bar{A} + B = \infty$ and $A\bar{B} = 0$, so from $F_{11} \prec F_{uv}$

we get $$\bar{F}_{uv} + F_{uv} = \infty,$$

and $$F_{11}\bar{F}_{11} = 0;$$

hence F_{11} and $\bar{F}_{11}$ are "contraries" of each other, and F_{uv}, $\bar{F}_{uv}$ are "sub-contraries." In the same way $F_{1v'} \prec F_{1v}$

gives $$\bar{F}_{v1} + F_{1v} = \infty,$$

and $$F_{1v'}\bar{F}_{v1} = 0;$$

that is, $F_{1v'}$ and $\bar{F}_{v'1}$ are contraries, and $\bar{F}_{v1}$, F_{1v} are sub-contraries. The line over F affects only F, not the suffices. Thus the negative of F_{11} would be written $\overline{(F_{11})}$, not $\bar{F}_{11}$.

To say "no U is F, during V" is evidently the same as to say "all U is $\bar{F}$, during V;" that is,

$$F_{01} = \bar{F}_{11};$$

$$\text{so } F_{10} = \bar{F}_{11},$$

$$\therefore F_{00} = F_{11}.$$

Since every proposition with zero as one or both of the suffices is thus expressible in a form with no suffix equal to zero, each suffix used will be supposed greater than zero. The suffices u, v are also supposed less than U, V, just as u was supposed less than U in the preceding section. $F_{\alpha\beta}$ will sometimes be used to include all six of the fundamental propositions: that is, α will be considered as having any one of the values 1, u, or u'; β as having any one of the values 1, v, or v'.

For *inference by combination* of such propositions we have the following simple rules, which are seen to be the same as in § 1:—

The conclusion from the *product* of two premises is the *product* of their predicates affected by suffices which are the *products* (in *extension*) of the suffices of the premises. Thus

$$F_{\alpha\beta} G_{\lambda\mu} \prec (FG)_{\alpha\lambda,\, \beta\mu}.$$

When *all* the suffices are 1, the relation between the *product* of the premises and the conclusion is equality; otherwise it is $\prec$,—that is, implication. Thus

$$F_{11} G_{11} = (FG)_{11},$$

$$F_{u1} G_{11} \prec (FG)_{u1},$$

etc.

The conclusion from the *sum* of two premises is the *sum* of their predicates affected by suffices which are the *sums* (in *intension*) of the suffices of the premises. Thus

$$F_{\alpha\beta} + G_{\lambda\mu} \prec (F + G)_{\alpha+\lambda,\, \beta+\mu}.$$

When *none* of the suffices are 1, the relation between the *sum* of the premises and the conclusion is equality; otherwise it is $\prec$,—that is, implication. Thus

$$F_{uv} + G_{uv} = (F + G)_{uv},$$

$$F_{1v} + G_{uv} \prec (F + G)_{uv},$$

etc.

But, by an exception to the rule, do *not* have $F_{u1} G_{1v} \prec (FG)_{uv}$ since G_{1v} is not of the form $(G_1)_v$.

These formulæ really follow at once from those in § 1. Thus F_{11} may be written $(F_1)_1$; hence by § 1 we have

$$(F_1)_1 (G_1)_1 = (F_1 G_1)_1 = \big((FG)_1\big)_1 = (FG)_{11},$$

and

$$(F_u)_v + (G_u)_v = (F_u + G_u)_v = \big((F + G)_u\big)_v = (F + G)_{uv}.$$

So in general we have

$$(F_\alpha)_\beta (G_\lambda)_\mu \prec (F_\alpha G_\lambda)_{(\beta\mu)} \prec (FG)_{\alpha\lambda,\, \beta\mu},$$

and

$$(F_\alpha)_\beta + (G_\lambda)_\mu \prec (F_\alpha + G_\lambda)_{\beta+\mu} \prec (F + G)_{\alpha+\lambda,\, \beta+\mu},$$

the addition of the suffices being taken in the same sense as in § 1; that is,

$$1 + 1 = 1, \qquad 1 + u' = u',$$
$$1 + u = u, \qquad u + u' = u,$$
$$u + u = u, \qquad u' + u' = u',$$

with like equations for v, v'. The second set of equations means

All of + same part of = same part of,
Some of + same part of = some part of,
Same part of + same part of = same part of,

and a little consideration will show that the formulæ hold as well for the accented suffices as for the unaccented.

The following formula is evident:—

$$(FG)_{\alpha\beta} \prec F_{\alpha\beta} G_{\alpha\beta}.$$

For *inference by elimination* we have only to consider the general form $F_{\alpha\beta}$, and the rule is precisely the same as the rule for elimination given in § 1, viz.: *Any set of terms may be eliminated by erasure provided no aggregant term is thereby destroyed.* Thus

$$(a + b\bar{x} + \bar{c}dxy + \bar{e}y)_{\alpha\beta} \prec (a + b + \bar{c}d + \bar{e})_{\alpha\beta},$$

and the reason of the rule need not be repeated.

The rule for *inference by predication* is also evidently the same as that previously given. Thus

$$(a + b + \bar{c}d + \bar{e})_{\alpha\beta} \prec (c \prec a + b + \bar{e})_{\alpha\beta},$$

and, in general,

$$(F)_{\alpha\beta} \prec (\ m = mF)_{\alpha\beta}.$$

If, after the multiplication has been performed, $mF = mP$, then we have

$$F_{\alpha\beta} \prec (m \prec P)_{\alpha\beta}.$$

Since propositions of the form F_{11} can be ***multiplied*** without loss of content, and propositions of the form F_{uv} can be *added* without loss of content, the most general proposition involving the six fundamental elements is of the form

$$\Sigma\,(F_{11}\Pi G_{u1}\Pi H_{u'1}\Pi K_{1v'}\Pi L_{1v}\Pi M_{uv}),$$

or $$\Pi\,(\Sigma F_{11} + \Sigma G_{u1} + \Sigma H_{u'1} + \Sigma K_{1v} + \Sigma L_{1v} + M_{uv}),$$

where F, G, etc. are logical polynomials of class terms. But to the six elements just considered we may add as elements the forms Φ_1, Φ_v considered at the close of § 1, where Φ is of the form $P_u + \Sigma Q_1$, or $P_1 \Pi Q_u$ (see page 79); so that Φ_1, Φ_v will be of the forms

$$(P_u + \Sigma Q_1)_1,\ (P_1 \Pi Q_u)_v.$$

It is clear that $(P_u + \Sigma Q_1)_v = P_{uv} + \Sigma Q_{1v}$, and that $(P_1 \Pi Q_u)_1 = P_{11} \Pi Q_{u1}$; but for the two forms of Φ_1, Φ_v just given, no such reduction can be made. The suffices within the parentheses of Φ_1, Φ_v refer to the universe of class terms, those outside to the universe of time. If the relative meaning of these suffices be reversed, so that the suffices inside the parentheses refer to the universe of class terms and those outside to the universe of time, we have two other propositional elements. Thus in order to distinguish the meaning of the suffices clearlý, it will be necessary to use the capital letters U, V, and write the four forms just considered as

$$\Phi_V,\ \Psi_v,\ X_U,\ \Omega_u,$$

or, in full,

$$(P_u + \Sigma Q_U)_V,\ (P_U \Pi Q_u)_v,\ (P_v + \Sigma Q_V)_U,\ (P_V \Pi Q_v)_u.$$

The negative of Φ_V is $\bar{\Phi}_v$, which is of the form Ψ_v. So the negative of X_U is $\bar{X}_u$, which is of the form Ω_u. As examples of Φ_V, X_U, suppose the universe of class terms to be plane figures a, b, etc., on a blackboard, and the universe of time to be an hour. Let $P = ab$, and $Q = \bar{c} + \bar{d}$; then

$$\{(ab)_u + (\bar{c} + \bar{d})_U\}_V$$

means "during every part of the hour, either some a is b, or no c is d," while

$$\{(ab)_v + (\bar{c} + \bar{d})_V\}_U$$

means "for every part of the blackboard, it is true that it is either sometimes both a and b, or never both c and d." So, as examples of Ψ_v and Ω_u we have, respectively,

$$\{(ab)_U(\bar{c} + \bar{d})_u\}_v,$$

which means "at some time during the hour, all the blackboard is ab, and some of it is $\bar{c} + \bar{d}$," and

$$\{(ab)_V(\bar{c} + \bar{d})_v\}_u,$$

which means "some part of the blackboard is always ab and sometimes $\bar{c} + \bar{d}$."

Adding the four propositional elements just described to the six described previously, we see that the most general proposition is of the form

$$\Pi(\Sigma F_{11} + \Sigma G_{u1} + \Sigma H_{u1} + \Sigma K_{1v} + \Sigma L_{1v} + M_{uv} + \Sigma\Phi_V + \Psi_v + \Sigma X_U + \Omega_u).$$

To illustrate the method of inference from propositions like the foregoing, consider the solution of the following problem: —

Six plane figures, a, b, c, d, e, f, *on a blackboard are constantly changing their size, shape, and position during an hour under the following restrictions:* —

I. *The area of* c *and* d *together is always included in the area of* a *and* b *together, or else, during a certain portion of the hour,* e *is equal to the part common to* d *and* f.

II. *The part of* a *which is not* e *is always included under the part common to* d *and* f *which is not* b, *or else, during the whole hour, it is true for some part of the board that all* b *is both* c *and* e.

III. *Either* a *and* d *are non-existent and* e *always covers the board, or else it is always covered either by* b *or by* c.

What may be inferred (1) *about the relation among* a, c, e *and* f, *independent of* b *and* d; (2) *about the relation among* a, c, e, *independent of* b, d, f?

The premises are

$$\text{I.}\qquad (a + b + \bar{c}\bar{d})_{11} + (def + \bar{d}\bar{e} + \bar{e}\bar{f})_{1v'}$$

$$\text{II.}\qquad (\bar{a} + e + \bar{b}df)_{11} + (\bar{b} + ce)_{u1}$$

$$\text{III.}\qquad (\bar{a}\bar{d}e)_{11} + \{(b)_U + (c)_U\}_{V}$$

From the product of the first two we infer

$$(\bar{a}b + \bar{a}\bar{c}\bar{d} + ae + be + \bar{c}\bar{d}e + a\bar{b}df)_{11} + (def + \bar{a}\bar{d}\bar{e} + \bar{a}\bar{e}\bar{f})_{1v'} \\ + (a\bar{b} + \bar{b}\bar{c}\bar{d} + ace + bce)_{u1} + (\bar{b}def + \bar{b}\bar{d}\bar{e} + \bar{b}\bar{e}\bar{f} + cdef)_{uv};$$

and multiplying this proposition by the third premise according to the preceding rules, we get as an inference

$$(\bar{a}b\bar{d}e + \bar{a}\bar{c}\bar{d}e)_{11} + (\bar{a}\bar{b}\bar{c}\bar{d}e + \bar{a}bc\bar{d}e)_{u1} + (bdef + \bar{a}b\bar{d}\bar{e} + \bar{a}b\bar{e}\bar{f})_{1v'} \\ + (cdef + \bar{a}c\bar{d}\bar{e} + \bar{a}c\bar{e}\bar{f})_{1v'} + (a\bar{b}c + ace + bce)_{u1} + (cdef + \bar{b}c\bar{d}\bar{e} \\ + \bar{b}c\bar{e}\bar{f})_{uv} + \{(\bar{a}b + be)_U + (\bar{a}bc + ace + bce + a\bar{b}cdf)_U\}_{V},$$

three of the complex elements reducing to simple ones according to the formulæ,

$$(F_U + G_U)_{v'} \prec F_{1v'} + G_{1v'},$$

$$(F_u + G_u)_V = (F + G)_{u1},$$

$$(F_u + G_u)_v = (F + G)_{uv}.$$

Dropping b and d from the above proposition, we get

$$(\bar{a}e)_{11} + (\bar{a}e)_{u1} + (\bar{a}\bar{e} + ef)_{1v'} + (\bar{a}c\bar{e} + cef)_{1v'} + (ac + ce)_{u1} \\ + (c\bar{e} + cf)_{uv} + \{(\hat{a} + e)_U + (\bar{a}c + ce + cf)_U\}_{V}.$$

But in a sum, any term may be dropped which implies, or is included under, another term.

$$(\bar{a}e)_{11} \prec (\bar{a}e)_{u1} \quad \text{and} \quad (\bar{a}c\bar{e} + cef)_{1v'} \prec (\bar{a}\bar{e} + ef)_{1v'};$$

therefore the above reduces to

$$(\bar{a}e)_{u1} + (\bar{a}\bar{e} + ef)_{1v'} + (ac + ce)_{u1} + (c\bar{e} + cf)_{uv} + \{(\bar{a} + e)_U + (\bar{a}c + ce + cf)_U\}_{V},$$

which is the first quaesitum, and may be read in words "either it is always true that some e is not a; or at a particular part of the hour all a is e, and all e is f; or during each part of the hour some c is either a or e; or at some part of the hour some c is either f or not e; or during each part of the hour either all a is e, or the whole blackboard is c and all a is either e or f."

Dropping f from this result, we get

$$(\bar{a}e)_{u1} + (\bar{a} + e)_{1v'} + (ac + ce)_{u1} + (c)_{uv} + \{(\bar{a} + e)_U + (c)_U\}_V.$$

But $(ac + ce)_{u1} \prec (c)_{uv}$ and $\{(\bar{a} + e)_U + (c)_U\}_V \prec (\bar{a} + e)_{1v'} + (c)_{uv}$, therefore we get as the second quæsitum,

$$(\bar{a}e)_{u1} + (\bar{a} + e)_{1v'} + (c)_{uv},$$

which means "either it is always true that some e is not a; or during some particular part of the hour all a is e; or there is sometimes some c." In like manner any other set of terms can be eliminated by dropping them from the product of the premises.

Propositions of more than two dimensions. If the universe of relation be supposed to consist of three dimensions, U, V, W, proceeding just as before we should find that the number of fundamental propositions with three suffices,

$$F_{111},\ F_{u11},\ F_{uv1},\ F_{uvw},\ \text{etc.},$$

is twenty-six. The logic of such propositions is a "hyper"

logic, somewhat analogous to the geometry of "hyper" space. In the same way the logic of a universe of relation of four or more dimensions could be considered. The rules of inference would be exactly similar to those already given.

Allusion has already been made to the fact that the propositions considered in this and the preceding section may be regarded as relative terms. In the first section, the two fundamental propositions, F_1 and F_u, are dual relatives. F_1 means "F is a *description of* every part of U;" and F_u means "F is a *description of* some part of U." Thus F_1 and F_u correspond to the two fundamental dual relatives. So in § 2, F_{11} is a triple relative term, meaning "F is a *description of* every part of U *during* every part of V." Thus the six fundamental propositions of two dimensions correspond exactly to the six fundamental varieties of triple relatives, and so on.

§ 3. *On Certain Other Methods.*

The propositions A and O in Mr. Peirce's notation are, respectively,

$$X \prec Y,$$

$$X \overline{\prec} Y.$$

Mr. McColl expresses them in a similar way, using a different symbol for the copula. Both Mr. McColl and Mr. Peirce have given algebraic methods in logic, in which the terms of these propositions are allowed to remain on both sides of the copula.

In the method of § 1 (of which § 2 is an extension), the propositions A and O are expressed as follows: —

$$(\bar{X} + Y)_1, \quad \text{equivalent to} \quad \infty \prec \bar{X} + Y,$$

$$(X\bar{Y})_u, \quad \text{"} \quad \text{"} \quad X\bar{Y} \overline{\prec} 0;$$

that is, all the terms of the universal proposition are transposed to the right hand side of the copula, while those of the particular proposition are transposed to the left-hand side.

If these propositions be expressed in the reverse way, namely, —

$$X\bar{Y} -\!\!< 0,$$

$$\infty \;\overline{-\!\!<}\; \bar{X} + Y,$$

the rules of inference become the exact logical negatives of those in § 1, addition taking the place of multiplication, and *vice versa*. $X\bar{Y} -\!\!< 0$ is equivalent to $(X\bar{Y})_0$, meaning "none of U is $X\bar{Y}$," as has already been explained. $\infty \;\overline{-\!\!<}\; \bar{X} + Y$ may be represented by $(\bar{X} + Y)_q$, meaning "some of U is not $\bar{X} + Y$," or "there is something besides $\bar{X} + Y$." Thus F_0 and F_q are the two fundamental forms of proposition in this method, and the rules of *inference by combination* are

$F_0 G_0 = (F + G)_0$	$F_q + G_q = (FG)_q$
$F_0 G_q -\!\!< (F + G)_q$	$F_q + G_0 -\!\!< (FG)_q$
$F_q G_q -\!\!< \infty.$	$F_0 + G_0 -\!\!< (FG)_0.$

Elimination is performed by *multiplying* together the co-efficients of the quantities to be eliminated.

Boole's method, as simplified by Schröder, has been extended by Miss Ladd, in the foregoing paper, so as to express particular propositions without the use of Boole's objectionable "arbitrary" class symbol. She has expressed A and O as follows: —

$$X\bar{Y}\,\overline{\vee}, \quad \text{equivalent to} \quad X\bar{Y} -\!\!< 0,$$

$$X\bar{Y}\,\vee, \qquad \text{"} \qquad \text{"} \quad X\bar{Y} \;\overline{-\!\!<}\; 0.$$

Thus F_0 and F_u are the two fundamental forms of proposition in her method, and the rules of *inference by combination* are

$F_0 G_0 = (F + G)_0$	$F_u + G_u = (F + G)_u$
$F_0 G_u \prec (\bar{F} G)_u$	$F_u + G_0 \prec (F + \bar{G})_u$
$F_u G_u \prec \infty.$	$F_0 + G_0 \prec (FG)_0.$

Elimination from F_0 is performed by *multiplying* coefficients; from F_u, by *adding* them.

One more method remains to be noticed,—the negative of Miss Ladd's method, in which A and O are expressed as

$$\infty \prec \bar{X} + Y,$$
$$\infty \overline{\prec} \bar{X} + Y,$$

and where F_1 and F_q are thus the two fundamental forms of proposition. The rules of *inference by combination* are

$F_1 G_1 = (FG)_1$	$F_q + G_q = (FG)_q$
$F_1 G_q \prec (\bar{F} + G)_q$	$F_q + G_1 \prec (F\bar{G})_q$
$F_q G_q \prec \infty,$	$F_1 + G_1 \prec (F + G)_1;$

and *elimination* from F_1 is performed by *addition* of coefficients; from F_q, by *multiplication* of coefficients.

§ 4. *On a special notation for De Morgan's Eight Propositions, with an extension of the same to similar propositions of three or more terms.*

It is proposed in this section so to change the notation previously given for De Morgan's eight propositions that the elimination of the middle term will be performed by an *algebraic* multiplication of the premises. Denote by I', E', O', A' what I, E, O, A become when each term is replaced by its negative. The propositions I, E, O, A, and their complementaries I', E', O', A' have already been represented (see page 76) respectively by

$(ab)_u$, $(\bar{a} + \bar{b})_1$, $(a\bar{b})_u$, $(\bar{a} + b)_1$, $(\bar{a}\bar{b})_u$, $(a + b)_1$, $(\bar{a}b)_u$, $(a + \bar{b})_1$;

and also, since $F_1 = \overline{F}_0$, by

$$(ab)_u,\ (ab)_0,\ (a\bar{b})_u,\ (a\bar{b})_0,\ (\bar{a}\bar{b})_u,\ (\bar{a}\bar{b})_0,\ (\bar{a}b)_u,\ (\bar{a}b)_0.$$

Let these be now changed to

$$(ab),(ab)^{-1},(ab^{-1}),(ab^{-1})^{-1},(a^{-1}b^{-1}),(a^{-1}b^{-1})^{-1},(a^{-1}b),(a^{-1}b)^{-1},$$

where the negative of a term is now denoted by affecting it with the exponent -1, and the negative of a proposition is denoted in the same way. Thus

(ab^{-1}) means "some a is not b,"

$(ab^{-1})^{-1}$ " "all a is b," etc.

With this notation there is the following simple

RULE OF INFERENCE. *Excluding products of two particulars, the conclusion from a set of premises is their algebraic product, with the convention that the appearance of a middle term in the result indicates that there is no conclusion.*

Thus, *Barbara* is

$$(mp^{-1})^{-1} \times (sm^{-1})^{-1} -\!\!< (sp^{-1})^{-1},$$

and *Darii* is

$$(mp^{-1})^{-1} \times (sm) -\!\!< (sp);$$

but from A and O as premises we get

$$(mp^{-1})^{-1} \times (sm^{-1}) -\!\!< \infty,$$

the middle term not disappearing from the product.

From the nature of this notation, just as with that of § 1, the order in which the two terms of a proposition are written is indifferent, and consequently the figure of a syllogism is indifferent. Thus, (mp) is the same as (pm). Thus *Celarent* and *Cesare* are

$$(mp)^{-1} \times (sm^{-1})^{-1} -\!\!< (sp)^{-1}.$$

Darii and *Datisi* are

$$(mp^{-1})^{-1} \times (sm) -\!\!< (sp).$$

Ferio, *Festino*, *Ferison*, and *Fresison* are

$$(mp)^{-1} \times (sm) -\!\!< (sp^{-1}).$$

Camestres and *Camenes* are

$$(pm^{-1})^{-1} \times (sm)^{-1} -\!\!< (sp)^{-1}.$$

Baroko is

$$(pm^{-1})^{-1} \times (sm^{-1}) -\!\!< (sp^{-1}).$$

Disamis and *Dimaris* are

$$(mp) \times (ms^{-1})^{-1} -\!\!< (sp).$$

Bokardo is

$$(mp^{-1}) \times (ms^{-1})^{-1} -\!\!< (sp^{-1}).$$

This rule of inference is seen to accord with the now recognized invalidity of the moods *Darapti*, *Felapton*, and *Fesapo*. Thus the premises of *Darapti* are

$$(mp^{-1})^{-1} \times (ms^{-1})^{-1},$$

from the product of which m does not disappear, and there is therefore, according to the rule, no inference. The same is true for *Felapton* and *Fesapo*. The premises of *Bramantip* are

$$(pm^{-1})^{-1} \times (ms^{-1})^{-1}, \text{ which } -\!\!< (s^{-1}p)^{-1}.$$

The following Table gives all the valid moods from De Morgan's eight propositions:—

	$(pm^{-1})^{-1}$	$(p^{-1}m)^{-1}$	$(p^{-1}m^{-1})^{-1}$	$(pm)^{-1}$	(pm^{-1})	$(p^{-1}m)$	$(p^{-1}m^{-1})$	(pm)
$(sm^{-1})^{-1}$		$(sp^{-1})^{-1}$		$(sp)^{-1}$	$(s^{-1}p)$		$(s^{-1}p^{-1})$	
$(s^{-1}m)^{-1}$	$(s^{-1}p)^{-1}$		$(s^{-1}p^{-1})^{-1}$			(sp^{-1})		(sp)
$(s^{-1}m^{-1})^{-1}$		$(s^{-1}p^{-1})^{-1}$		$(s^{-1}p)^{-1}$	(sp)		(sp^{-1})	
$(sm)^{-1}$	$(sp)^{-1}$		$(sp^{-1})^{-1}$			$(s^{-1}p^{-1})$		$(s^{-1}p)$
(sm^{-1})	(sp^{-1})		(sp)					
$(s^{-1}m)$		$(s^{-1}p)$		$(s^{-1}p^{-1})$				
$(s^{-1}m^{-1})$	$(s^{-1}p^{-1})$		$(s^{-1}p)$					
(sm)		(sp)		(sp^{-1})				

There are twenty-four valid moods, but if no distinction be made between s and p, these reduce to the twelve in either half of the Table, the Table dividing itself symmetrically along the diagonal from left down to right. The unsymmetry of the Aristotelian system is seen from the fact that the fifteen valid moods of the Aristotelian system comprise only eight out of the twenty-four of the Table, and these eight select themselves very unsymmetrically, being those underscored by dotted lines.

From the three formulæ

$$(sm^{-1})^{-1} \times (pm)^{-1} \prec (sp)^{-1},$$

$$(sm) \times (pm)^{-1} \prec (sp^{-1}),$$

$$(sm)^{-1} \times (pm) \prec (s^{-1}p),$$

the whole twenty-four syllogisms of the Table may be obtained by substituting for m, s, and p their negatives in all possible ways, each formula yielding eight.

Mr. Hugh McColl, in his papers on logic in the "Proceedings of the London Mathematical Society" (Vol. IX, *et. seq.*), has been using a notation for the copula identical in meaning with that of Mr. Peirce. He uses a colon to denote implication, instead of $\prec$. Mr. Peirce has recently told me that Mr. McColl justifies his use of the colon by its mathematical meaning as a sign of division. Thus *Barbara* and *Celarent* are

$$\begin{array}{ll} m : p & m : \bar{p} \\ s : m & s : m \\ \therefore s : p & \therefore s : \bar{p}, \end{array}$$

and the analogy to division is obvious. But this analogy

exists only in the two universal moods of the first figure. Thus *Cesare* and *Festino* are

$$p : \bar{m} \qquad\qquad p : \bar{m}$$

$$s : m \qquad\qquad s \div \bar{m}$$

$$\therefore\ s : \bar{p} \qquad\qquad s \div p,$$

where $\div$ is the negative copula, and the analogy to division is wanting. In the notation of this section the analogy of the premises to ratios, and of the conclusion to their product is more nearly complete.

Extension of the preceding.

Let (abc) denote "a, b, c have something in common,"

and $(abc)^{-1}$ " a, b, c " nothing " "

By substituting for a, b, c their negatives in all possible ways, we get sixteen propositions concerning three terms, thus seen to be analogous to De Morgan's eight concerning two terms. In the same way we may get thirty-two propositions concerning four terms, and 2.2^n propositions concerning n terms. The formulæ of inference from propositions like the above are

$$(ab\ldots gh\ldots l)\ (h\ldots lm\ldots q)^{-1} \prec (ab\ldots g)\ (m\ldots q)^{-1},$$

$$(ab\ldots\ldots kl)\ (l^{-1}m\ldots\ldots q)^{-1} \prec (ab\ldots\ldots km\ldots q)^{-1}.$$

In the first, where one premise is particular, inference can take place independently of any number of middle terms, provided each term is positive in both premises, or negative in both. In the second formula, when both premises are universal, inference can take place independently of only one middle term, and this must be of different quality in the two premises. By an obvious substitution these two formulæ are reduced to the formulæ previously given involving only two terms in each premise. Thus

$$(xy) \quad (yz)^{-1} \prec (xz^{-1}),$$

$$(xy)^{-1} \; (y^{-1}z)^{-1} \prec (xz)^{-1}.$$

That is, the premises of the first mean "that which is common (x) to $a, b, \ldots g$, has something in common with the common part (y) of $h, \ldots l$;" and "the common part (y) of $h, \ldots l$ has nothing in common with $m, \ldots q$." Whence the inference is (xy^{-1}), or $(ab \ldots g)\,(m \ldots q)^{-1}$. The premises of the second mean "whatever may be common (x) to $a, b, \ldots k$, has nothing in common with l;" and "whatever may be common (z) to $m, \ldots q$, has nothing in common with non-l." Whence the inference is $(xz)^{-1}$, or $(ab \ldots km \ldots q)^{-1}$.

$$(abc) \text{ means } (ab)\;(ac)\;(bc),$$

$$\therefore\; (abc)^{-1} \text{ `` } \quad (ab)^{-1} + (ac)^{-1} + (bc)^{-1}.$$

Thus any one of these propositions is reducible to a function of De Morgan's eight.

§ 5. *Note on De Morgan's Twenty Propositions.*[1]

It is proposed in this section to consider a simple method of deriving and writing De Morgan's Twenty Propositions. Let A = all of A, a = part of A, $\overline{A}$ = all of non-A, and $\bar{a}$ = part of non-A, where *part of* is understood to mean *less than the whole of.* Let a second term B be modified in the same way. Then, by affirming and denying identity between each modification of the first term and each modification of the second, we get thirty-two propositions, of which, however, twelve are duplicates. That is, the process yields twenty distinct

[1] See his "Syllabus of Logic," §§ 24–62.

propositions, and they are easily seen to be the twenty of De Morgan. Let the affirmation of identity between two terms be denoted by their juxtaposition, and let the denial of the same be denoted by a line extending over both terms. Then we have the following

TABLE OF DE MORGAN'S TWENTY PROPOSITIONS.

AB, or $\bar{A}\bar{B}$	$\bar{A}B$, or $A\bar{B}$	$\overline{AB}$, or $\overline{\bar{A}\bar{B}}$	$\overline{\bar{A}B}$, or $\overline{A\bar{B}}$
Ab, " $\bar{a}\bar{B}$	$\bar{A}b$, " $a\bar{B}$	$\overline{Ab}$, " $\overline{\bar{a}\bar{B}}$	$\overline{\bar{A}b}$, " $\overline{a\bar{B}}$
aB, " $\bar{A}\bar{b}$	$\bar{a}B$, " $A\bar{b}$	$\overline{aB}$, " $\overline{\bar{A}\bar{b}}$	$\overline{\bar{a}B}$, " $\overline{A\bar{b}}$
ab	$\bar{a}b$	$\overline{ab}$	$\overline{\bar{a}b}$
$\bar{a}\bar{b}$	$a\bar{b}$	$\overline{\bar{a}\bar{b}}$	$\overline{a\bar{b}}$

Thus, AB means "the whole of A is identical with the whole of B." It is obvious that $\bar{A}\bar{B}$ is equivalent in meaning to AB. The second proposition, Ab, means "the whole of A is identical with a part of B" (that is, all A is B, and some B is not A). It is clear that $\bar{a}\bar{B}$, or "a part of non-A is identical with the whole of non-B," is the same as Ab. To take an example from the other side of the Table, $\overline{AB}$ means "it is not true that the whole of A is identical with the whole of B." This is simply the denial of the proposition AB. $\overline{Ab}$ means "it is not true that the whole of A is identical with a part of b," a simple denial of Ab.

The propositions below the horizontal line of division, which are differentiated from those above the line by containing only small letters in their symbols, are De Morgan's eight "simple" propositions.

$$ab = \left\{\begin{array}{l}\text{A part of } A \text{ is a part of } B \\ \text{Some } A \text{ is } B\end{array}\right\} \quad . \quad . \quad . \quad . \quad . \quad . \quad . \quad . \quad I.$$

$$\overline{ab} = \left\{\begin{array}{l}\text{It is not true that a part of } A \text{ is a part of } B \\ \text{No } A \text{ is } B\end{array}\right\} \quad . \quad . \quad E.$$

$$a\bar{b} = \left\{\begin{array}{l}\text{A part of } A \text{ is a part of non-}B \\ \text{Some } A \text{ is not } B\end{array}\right\} \quad . \quad . \quad . \quad . \quad . \quad . \quad . \quad O.$$

$$\overline{a\bar{b}} = \left\{\begin{array}{l}\text{It is not true that a part of } A \text{ is a part of non-}B \\ \text{All } A \text{ is } B\end{array}\right\} \quad . \quad A.$$

The remaining four of these eight are derived from these four by the negation of their terms. This notation for the eight propositions differs only slightly from that employed in previous sections.

De Morgan derived his eight "simple" propositions by applying the Aristotelian forms A, E, I, O to the four pairs of terms X, Y; $X, \bar{Y}$; $\bar{X}, Y$; $\bar{X}, \bar{Y}$. This process gives sixteen propositions, of which eight are duplicates. The other twelve of the twenty he called "complex," because they are compounded of the eight simple propositions, as follows: —

$$\begin{array}{ll}
AB = \overline{a\bar{b}} \times \overline{\bar{a}b} & \qquad \overline{AB} = a\bar{b} + \bar{a}b \\
Ab = \overline{a\bar{b}} \times \bar{a}b & \qquad \overline{Ab} = a\bar{b} + \overline{\bar{a}b} \\
aB = a\bar{b} \times \overline{\bar{a}b} & \qquad \overline{aB} = \overline{a\bar{b}} + \bar{a}b \\
\bar{A}B = \overline{\bar{a}\bar{b}} \times \overline{ab} & \qquad \overline{\bar{A}B} = \bar{a}\bar{b} + ab \\
\bar{A}b = \overline{\bar{a}\bar{b}} \times ab & \qquad \overline{\bar{A}b} = \bar{a}\bar{b} + \overline{ab} \\
\bar{a}B = \bar{a}b \times \overline{ab} & \qquad \overline{\bar{a}B} = \overline{\bar{a}\bar{b}} + ab
\end{array}$$

The following Table gives the conclusions from one hundred out of the possible four hundred combinations of two premises from this system of twenty propositions:

	PM	Pm	pM	pm	$\bar{p}\bar{m}$	$\overline{PM}$	$\overline{Pm}$	$\overline{pM}$	$\overline{pm}$	$\overline{\bar{p}\bar{m}}$
SM	SP	sP	Sp	sp	$\bar{s}\bar{p}$	$\overline{SP}$	$\overline{sP}$	$\overline{Sp}$	$\overline{sp}$	$\overline{\bar{s}\bar{p}}$
Sm	Sp	$\bar{s}\bar{p}$	Sp	o	$\bar{s}\bar{p}$	o	o	o	$S\bar{p}$	$\bar{s}p$
sM	sP	sP	sp	sp	o	o	o	o	$s\bar{p}$	$s\bar{P}$
sm	sp	o	sp	o	o	o	o	o	$s\bar{p}$	o
$\bar{s}\bar{m}$	$\bar{s}\bar{p}$	$\bar{s}\bar{p}$	o	o	o	o	o	o	o	$\bar{s}p$
$\overline{SM}$	$\overline{SP}$	o	o	o	o	o	o	o	o	o
$\overline{Sm}$	$\overline{Sp}$	o	o	o	o	o	o	o	o	o
$\overline{sM}$	$\overline{sP}$	o	o	o	o	o	o	o	o	o
$\overline{sm}$	$\overline{sp}$	$\bar{s}P$	$\bar{s}p$	$\bar{s}p$	o	o	o	o	o	$\overline{s\bar{p}}$
$\overline{\bar{s}\bar{m}}$	$\overline{\bar{s}\bar{p}}$	$s\bar{p}$	$\bar{S}p$	o	$s\bar{p}$	o	o	o	$\bar{\bar{s}}p$	o

By applying the sign of negation first to the S, then to the P, then to both the S and the P, the remaining three hundred are obtained. According to De Morgan, who postulates that every term and its negative is greater than zero, there are two conclusions not given in the Table, namely:—

$$\overline{sm} \times \overline{pm} \prec \bar{s}\bar{p},$$

$$\overline{\bar{s}\bar{m}} \times \overline{\bar{p}\bar{m}} \prec sp,$$

and from these are obtained six others by applying the sign of negation to s and p. But according to the definitions of Mr. Peirce and others, already alluded to, these are invalid conclusions; since, being particular, they imply the existence of their subjects, while the universal premises do not.

OPERATIONS IN RELATIVE NUMBER WITH APPLICATIONS TO THE THEORY OF PROBABILITIES.

By B. I. Gilman.

The purpose of this Paper is to deduce the formulæ for the addition and multiplication of Relative Number, and to apply them in demonstrating the well-known fundamental theorems of Probabilities, according to Mr. Peirce's method of dealing with the subject.

If a relation be that which we perceive when a group of objects are viewed together, but which we do not perceive when we regard each separately, then any act of comparison will bring to view a relation. If the objects compared are two in number, the relation may be called a dual one.

Such a dual relation may be viewed in two lights, or we may say it splits into two elementary forms, according as one or the other object is our starting-point in comparing the couple. The two are called the direct relation and its converse. Thus, what is ordinarily termed a relation may be said to have ends, being based on a comparison having a direction. One of these ends is called the *relate*, the other the *correlate*.

A relative number is a number obtained in either of the two following ways: first, by dividing the number of instances in which a given relation has a relate in a certain class of objects by the number of objects in the class; or, second, by dividing the number of instances in which a given relation has a correlate in the given class by the number of objects in the class. Hence, for a given relation ρ' we have two such relative or average numbers, — one, the number of instances in which ρ' has a relate of the class y, divided by the number of y's; and the other, the average number per y of ρ' whose correlates are y's. The former might be called the relate-number of ρ', the latter its correlate-number. But if we extend the class y to include all the objects in the universe, since the number of instances in which the relation ρ' occurs having a relate which is an object in the universe, is equal to the total number of times ρ' occurs at all, and the same thing is true of the number of occurrences in which it has a correlate which is in the universe: it follows that for both relate and correlate numbers we get the average number of relations ρ' per object in the universe. That is, any relation ρ' has but one (what we shall call) general relative number.

Denoting each object in the universe by a certain letter, each possible different couple of objects (considering those couples as different in which the same elements occur in a different order) will be symbolized once, and only once, in Mr. Peirce's scheme of pairs, as follows: —

A : A	A : B	A : C	A : D	. . .
B : A	B : B	B : C	B : D	. . .
C : A	C : B	C : C	C : D	. . .
D : A	D : B	D : C	D : D	. . .
. . .	. . .	. . .	. . .	. . .

Now if in this scheme of pairs we assume the relation-direction to be constant, say from left to right, — that is, that the right-hand members of the pairs are the correlates, — it will follow that any single instance of any relation must subsist between some one, and only one, of the pairs. Marking in any way, as by a circumscribed circle, those pairs between the components of which subsists the relation ρ'; and marking by a circumscribed square instances of the relation ρ'', — we shall have in general some pairs surrounded by circles, some by squares, and some by both.

Whence if ρ' and ρ'' denote respectively the number of individual relations comprised in the general relations ρ' and ρ'', we shall have

$\rho' + \rho'' =$ number of pairs surrounded by circle alone + number of pairs surrounded by square alone + twice the number of pairs surrounded by both circle and square $= \rho', \bar{\rho}'' + \rho'', \bar{\rho}' + 2\,\rho', \rho''$

in which ρ', ρ'' denotes the number of pairs concerning each of which it can be said that it is in both the relations ρ' and ρ''; and $\rho', \bar{\rho}''$ denotes the number of pairs which are at once in the relation ρ' and not in the relation ρ''. Again,

$\rho' + \rho'' =$ number of pairs in circle, or square, or both + number in both $= (\rho' \cdot\!|\!\cdot \rho'') + \rho', \rho''$,

in which — according to Mr. Jevons's notation — $(\rho' \cdot\!|\!\cdot \rho'')$ denotes that class of pairs concerning each member of which it can be said that it is either an instance of ρ' or of ρ'' or of both. Now, since a general relative number is the total number of individual instances of a relation, divided by the number of objects in the universe, if we indicate the number of objects in the universe by ∞, $\frac{\rho'}{\infty}$ will indicate the general relative number of the relation ρ'. Symbolizing this quotient by $[\rho']$, and dividing both sides of the above equations by ∞, we have

$$[\rho'] + [\rho''] = [\rho', \bar\rho''] + [\rho'', \bar\rho'] + 2[\rho', \rho''] = [\rho' \cdot|\cdot \rho''] + [\rho', \rho''].$$

We thus have reached two formulæ for the addition of two relative numbers. Similarly, we have for the addition of three relative numbers

$$\begin{aligned}[\rho'] + [\rho''] + [\rho'''] = {} & [\rho', \bar\rho'', \bar\rho'''] + [\rho'', \bar\rho', \bar\rho'''] + [\rho''', \bar\rho', \bar\rho''] \\ & + 2[\rho', \rho'', \bar\rho'''] + 2[\rho', \rho''', \bar\rho''] + 2[\rho'', \rho''', \bar\rho'] \\ & + 3[\rho', \rho'', \rho'''].\end{aligned}$$

or

$$\begin{aligned} = {} & [\rho' \cdot|\cdot \rho'' \cdot|\cdot \rho'''] + [\rho', \rho'', \bar\rho'''] + [\rho', \rho''', \bar\rho''] \\ & + [\rho'', \rho''', \bar\rho'] + 2[\rho', \rho'', \rho'''].\end{aligned}$$

Similar formulæ may be deduced for the addition of n relative numbers, as follows:—

$$\begin{aligned}& [\rho'] + [\rho''] + [\rho'''] + \ldots [\rho^n] \\ & = [\rho', \bar\rho'' \ldots \bar\rho^n] + [\rho'', \bar\rho', \bar\rho''' \ldots \bar\rho^n] + \ldots + [\rho^n, \bar\rho' \ldots \bar\rho^{n-1}] \\ & + 2([\rho', \rho'', \bar\rho''' \ldots \bar\rho^n] + \ldots + [\rho^{n-1}, \rho^n, \bar\rho' \ldots \bar\rho^{n-2}]) \\ & + 3([\rho', \rho'', \rho''', \bar\rho^{\text{iv}} \ldots \bar\rho^n] + \ldots + [\rho^{n-2}, \rho^{n-1}, \rho^n, \bar\rho' \ldots \bar\rho^{n-3}]) \\ & + \ldots \\ & + n[\rho'\rho'' \ldots \rho^n],\end{aligned}$$

or

$$\begin{aligned}& = [\rho' \cdot|\cdot \rho'' \cdot|\cdot \rho''' \ldots \cdot|\cdot \rho^n] \\ & + [\rho', \rho'', \bar\rho''' \ldots \bar\rho^n] + \ldots + [\rho^{n-1}, \rho^n, \bar\rho' \ldots \bar\rho^{n-2}] \\ & + 2([\rho', \rho'', \rho''', \bar\rho^{\text{iv}} \ldots \bar\rho^n] + \ldots + [\rho^{n-2}, \rho^{n-1}, \rho^n, \bar\rho' \ldots \bar\rho^{n-3}] \\ & + \ldots \\ & + (n-1)[\rho', \rho'' \ldots \rho^n].\end{aligned}$$

This latter formula gives, when the relations are mutually incompatible,—that is, when no two of them can subsist between the same pair,—a much simpler result:—

$$[\rho'] + [\rho''] + \ldots + [\rho^n] = [\rho' \cdot|\cdot \rho'' \cdot|\cdot \rho''' \ldots \cdot|\cdot \rho^n],$$

all the other terms reducing to zero.

To obtain a formula for the multiplication of relative numbers we notice that

$$\frac{x}{\rho''} \times \frac{\rho''}{\infty} = \frac{x}{\infty}, \quad \text{or} \quad \frac{x}{\rho''} \times [\rho''] = [x].$$

Let x, which may be any number, signify the number of different existing groups of three objects, such that the first is to the second in the relation ρ' and the second to the third in the relation ρ''. Such a group may be called a relative sequence, and may be denoted by $\rho'\rho''$ without the comma. Then

$$\frac{\rho'\rho''}{\rho''} \times [\rho''] = [\rho'\rho''].$$

If now

$$\frac{\rho'\rho''}{\rho''} = \frac{\rho'}{\infty},$$

the formula becomes

$$[\rho'] \, [\rho''] = [\rho'\rho''].$$

In this case, therefore, the product of the relative numbers of the two given relations equals the relative number of the sequence formed from them.

Multiplying numerator and denominator of $\frac{\rho'}{\infty}$ by the number of objects in the universe, it becomes $\frac{\rho' \times \infty}{\infty^2}$. The numerator of this fraction is a number equal to the number of different triplets obtained by combining each ρ' with every object in the universe. Between the second and third members of these triplets either the relation ρ' or $\bar{\rho}'$ must hold; and no relative sequence of the form $\rho'\rho''$ or $\rho'\bar{\rho}''$ can exist which does not appear among them. Hence the number $\rho' \times \infty$ equals the sum of the numbers of $\rho'\rho''$ and $\rho'\bar{\rho}''$. The denominator being the square of the number of objects in the universe is equal to the number of possible pairs, and each of these is either ρ'' or $\bar{\rho}''$. Hence

$$\frac{\rho'\infty}{\infty^2} = \frac{\rho'\rho'' + \rho'\bar{\rho}''}{\rho'' + \bar{\rho}''}$$

and

$$\frac{\rho'\rho''}{\rho''} = \frac{\rho'\rho'' + \rho'\bar{\rho}''}{\rho'' + \bar{\rho}''}$$

or

$$\frac{\rho'\rho''}{\rho''} = \frac{\rho'\bar{\rho}''}{\bar{\rho}''}$$

That is, the average number of sequences $\rho'\rho''$ per each ρ'' is the same as the average number of sequences $\rho'\bar{\rho}''$ per each $\bar{\rho}''$. Hence, whether the relations in which any given individual stands to the others in the universe are all $\bar{\rho}''$, or one or more ρ'' and the rest $\bar{\rho}''$, will make no difference on the average in the number of relative sequences whose first member is ρ' of which it is the intermediary. The number of such sequences in the case of any individual being the number of the objects standing to it in the relation ρ' multiplied by the number of objects in the universe, it follows that the number of objects standing to any given individual in the relation ρ' is not affected by the circumstance of its being ρ'' to one or more objects.

Similarly, from $\frac{\rho'\rho''}{\rho''} = \frac{\rho'}{\infty}$ we may get $\frac{\rho'\rho''}{\rho'} = \frac{\rho''}{\infty}$, whence $\frac{\rho'\rho''}{\rho'} = \frac{\rho'\rho'' + \bar{\rho}'\rho'}{\rho' + \bar{\rho}'}$ or $\frac{\rho'\rho''}{\rho'} = \frac{\bar{\rho}'\rho''}{\bar{\rho}'}$; that is, whether an object is correlate in any relations ρ' or not, will make no difference on the average in the number of ρ''s of which it is the relate.

For instance, letting ρ' indicate the relation borrower from, and ρ'' the relation trustee of, this condition expresses, first, the fact that a man's being a trustee makes no difference on the average in the number of borrowers from him; and, second, that a man's being a lender or not makes no difference on the average in the number of funds which he controls as trustee. Such relations, from one of which nothing can be inferred regarding the presence of the other, are called independent relations. Hence for independent relations,

$$[\rho'] \times [\rho''] = [\rho'\rho''].$$

The expression $\rho'\rho''$ here denoting the number of relative sequences of that form, if we define a compound relation to be a combination of such relative sequences as have the same individual object as relate, $'$, and also the same individual object as correlate, $''$, we shall have each compound relation consisting of as many sequences as it has intermediary objects. Hence, in order to express the number of $\rho'\rho''$s in terms of compound relations of that form, to the total number of compound relations we shall have to add the number of those which have two intermediaries, since they each contribute an extra sequence; and to this sum we must further add twice the number of compound relations having three intermediaries, three times those having four, etc. Hence we have for the number of relative sequences expressed in terms of compound relations,

$$\rho'\rho'' = \overset{\frac{\circ}{\circ}}{P'}P'' + \overset{2}{P'}P'' + 2\,\overset{3}{P'}P'' + \ldots (n-1)\,\overset{n}{P'}P''$$

wherein $\overset{\frac{\circ}{\circ}}{P'}P''$ denotes the total number of compound relations of the form $\rho'\rho''$ having whatever number of intermediaries; $\overset{2}{P'}P''$ denotes the number of such compound relations having two intermediaries, etc. Whence, dividing through by ∞, we have

$$[\rho'\rho''] = [\overset{\frac{\circ}{\circ}}{P'}P''] + [\overset{2}{P'}P''] + \ldots (n-1)\,[\overset{n}{P'}P''],$$

and the following formula results for the multiplication of independent relative numbers: —

$$[\rho'][\rho''] = [P' \overset{\frac{0}{0}}{} P''] + [P' \overset{2}{} P''] + 2[P' \overset{3}{} P''] \ldots (n-1)[P' \overset{n}{} P''].$$

By a somewhat different and a longer process of proof, it can be shown that for independent relations the following formula holds for the multiplication of n relative numbers: —

$$[\rho'][\rho''] \ldots [\rho^n] = [P' \overset{\frac{0}{0}}{\ldots} P^n] + [P' \overset{2}{\ldots} P^n]$$
$$+ 2[P' \overset{3}{\ldots} P^n] + \ldots + (v-1)[P' \overset{v}{\ldots} P^n]$$
$$+ \cdot \quad \cdot \quad \cdot \quad \cdot \quad \cdot \quad \cdot \quad \cdot \quad \cdot \quad \cdot$$

Here it is to be noted that the superscribed numbers do not refer to the number of intermediaries, but to the degree of connection, the number of ways in which relate $'$ and correlate ^{n} are connected by chains of relation.

The continued product of the numbers indicating the simultaneous intermediaries at the successive steps, it is easily seen, cannot be less than r nor greater than $r^{(n-1)}$, when the connection in the given relation is an r-fold one. Since permuting the multipliers does not change the left-hand member, the right-hand member remains constant in whatever order the elementary relatives are compounded.

Through the addition formula we have reached what we may call polynomial relative numbers, of the form $[\rho' +\!\!\!, \rho'' +\!\!\!, \ldots +\!\!\!, \rho^n]$ which expresses the relative number of that class of pairs, each one of which is an instance of some one or more of the relations $\rho' \ldots \rho^n$. In the case of incompatible relations we have the equation

$$[\rho' +\!\!\!, \rho'' +\!\!\!, \ldots +\!\!\!, \rho^n] = [\rho'] + [\rho''] + \ldots + [\rho^n].$$

Whence the multiplication of polynomial relative numbers reduces in the case of incompatible relations to that of monomials.

The involution of a monomial relative number gives the ordinary result of multiplication, except that all the elements of the resulting compound relation are the same. If we involve an incompatible polynomial, we shall get a result according to the multinomial theorem, consisting of monomial powers and products.

In order to apply these results to the theory of probabilities, we shall require to make a supposition in regard to the character of the relations we are to consider. If a relation is perceived whenever we compare objects, it follows that a relation will be noticed when we think of an object as existing at successive times; for this involves a comparison between its aspect at one time and at another.

This relation between objects which differ, so far as we see, only in existing at different times, we call identity. The pairs in the principal diagonal of the relative scheme exist in this relation only, since what we call the same or an identical object is both correlate and relate.

The relative number of the relation of identity is evidently unity, since it occurs once, and no more, for every individual in the universe. Now we can, if we please, agree to bring the various individual relations, — that is, relations subsisting between individual objects, — which together make up the total extension of the general relation identity, into various classes according to the character of the objects they identify. This will create as many kinds of relation of identity as there are classes of objects in the universe, and their relative numbers will vary from $\frac{1}{\infty}$ up to unity, and will express the proportion of objects of the different kinds in the universe.

Further, we may agree to take for the divisor of our relative number, for our y, instead of all the objects in the universe, some limited portion of them, say the class *b*. This will be a return to the special relative number mentioned at the beginning of the paper; but it is evident that since the relation whose relative number we seek is a relation of identity, every instance of it which has its relate in the class *b* will also have its correlate in that class, and *vice versa;* so that the relate and correlate number of the relation will be the same, and may be called simply its relative number. Such a relative number will mean the number of identity relations of the form *a* to be found among the relations pertaining to the individuals of the class *b* divided by the number of those individuals; that is, the number of *a*'s among the *b*'s, divided by the number of *b*'s, or, in other words, the proportion of the genus *b* that is of the species *a*. If we regard events as the objects between which the relations we are considering subsist, an identical relative number will express the proportion in which a certain species of event exists in a genus. With this ratio will vary the expectation with which we shall look to see a case of the genus a case also of the species; it may be said to measure the value of the genus as a proof of the species,—to measure, that is, the prove-ability, or probability, of the species from the standpoint of the genus.

On this view of probability it has to do, not with individual events, but with classes of events; and not with one class, but with a pair of classes,—the one containing, the other contained. The latter being the one with which we are principally concerned, we speak, by an ellipsis, of its probability without mentioning the containing class; but in reality probability is a ratio, and to define it we must have both correlates given.

An identical relative number, then, when the identities considered are events, will be the ratio of a specific to a generic occurrence; and this ratio is called the probability of the species with respect to the genus. The mathematical combination of probabilities will therefore take place in accordance with the formulæ for relative number already reached, with such modifications as result from their application to relations of identity.

In establishing by these formulæ the fundamental theorems of probabilities, let the individuals in the universe we are considering be events; and let a denote a certain kind of relation of identity between them,—that is, a certain class of events,—and $\bar{a}$ the remaining relations of identity, that is, all the rest of the events in the universe. The general relative numbers of a and $\bar{a}$—that is, the general probabilities of a and $\bar{a}$ in the universe—will be denoted by $[a]$ and $[\bar{a}]$.

From the addition formula we have

$$[a] + [\bar{a}] = [a \cdot\!|\!\cdot \bar{a}] + [a,\bar{a}].$$

The first term of the right-hand member is the relative number of that class of pairs, each of which exhibits either or both of the relations a and $\bar{a}$; and the second term of the right-hand member is the relative number of that class of pairs, each of which exhibits both the relations a and $\bar{a}$. But since by definition a is a part and $\bar{a}$ the rest of the existing relations of identity, no event exhibits them both, and $[a,\bar{a}] = 0$; while the number of relations $a \cdot\!|\!\cdot \bar{a}$ equals ∞, and hence $[a \cdot\!|\!\cdot \bar{a}] = 1$. Thus we have

$$[a] + [\bar{a}] = 1$$
$$[\bar{a}] = 1 - [a] \qquad (1)$$

or, the probability of the negative of an event equals unity minus the probability of the event.

The relations a and $\bar{a}$ are incompatible relations; that is, they cannot subsist at once between the same pair.

Incompatibility means, therefore, in the case of relations of identity between events, that no one event can be of both species; the species are mutually exclusive, — the events, as we say, cannot happen together. Such events may be called exclusives, and we may denote by the term alternatives specific events which together make up a genus; that is, exclusives one or other of which must happen if the generic event happen at all. The generic event consisting of the occurrence of any one of a number of exclusives may be called an alternating event.

The abridged form of the addition formula, when the relations are incompatible, gives the following as the probability of an alternating event: —

$$[a \cdot|\cdot b \cdot|\cdot c \cdot|\cdot \ldots \cdot|\cdot n] = [a] + [b] + [c] + \ldots + [n] \quad (2)$$

That is, the probability of an alternating event is equal to the sum of the probabilities of the exclusives of which it is composed.

The expression $a, b, c, \bar{d} \ldots \bar{n}$ denotes an event which is at once a, b, c, not $d \ldots$ and not n; and $[a, b, c, \bar{d} \ldots \bar{n}]$ denotes the probability of such a compound event. If we have certain events of known probability, $a, b, c \ldots n$ which are not exclusives, and wish to obtain the probability of the occurrence of some one, and only one, of them, the desired expression reduces to a sum of such compound probabilities. For the event in question will be either $(a, \bar{b}, \ldots \bar{m}, \bar{n})$, or $(\bar{a}, b \ldots \bar{m}, \bar{n})$, etc., or $(\bar{a}, \bar{b} \ldots \bar{m}, n)$; and these compounds being mutually exclusive, the event is an alternating one, and its probability is expressed as follows: —

$$[a, \bar{b} \ldots \bar{n} \cdot|\cdot \bar{a}, b \ldots \bar{n} \cdot|\cdot \ldots \cdot|\cdot \bar{a} \ldots \bar{m}, n] =$$
$$[a, \bar{b} \ldots \bar{n}] + [\bar{a}, b \ldots n] + \ldots + [\bar{a} \ldots \bar{m}, n]$$

This result being in terms of the probability of compound

events, to make it available we must have means of calculating compound probabilities from simple ones.

The formula obtained above for multiplying relative numbers expresses the result of such a multiplication in terms of the relative numbers of compound relations. In the case of identical relations, these would be compound relations of identity. But since no object or event is in the relation of identity to more than one object or event, — that is, itself, — each compound relation of identity must consist of a single relative sequence; accordingly all the terms after the first in the right-hand member of the multiplication formula disappear, the remaining term being the relative number of a relation of identity compounded of all the multiplied factors. But since all the objects concerned in this compound relation from relate ′ to correlate ″ are one and the same, it is no longer a sequence of relations, but a coexistence of special identities, — a coexistence of characters; and its relative number is the relative number of such coexistences, — of objects or events in which coexist all the given special identities that belong at once to all the given species. The condition that the relations should be independent, that is, that between any two of them,

$$\frac{a,b}{b} = \frac{a,\bar{b}}{\bar{b}}$$

for relations of identity becomes the condition that the proportion of b's that are also a's should equal the proportion of $\bar{b}$'s that are also a's; in other words, that an event is b should make it neither more nor less likely that it is also a case of a, and *vice versa.*

We thus see that the multiplication of identical relative numbers, when the relations are independent, will give the relative number of the events in which all the multiplied identities coexist. The probability of a compound event, therefore, when the components are independent, may be found by multiplying together the probabilities of all the components. Applying this principle to the case of the compound events

$$[a, \bar{b} \ldots \bar{n}] + [\bar{a}, b, \ldots \bar{n}] \ldots + [\bar{a} \ldots \bar{m}, n],$$

we have for the probability of the occurrence of one, and only one, of n independent non-exclusive events,

$$\begin{aligned}&[a, \bar{b} \ldots \bar{n} \cdot\!|\!\cdot \bar{a}, b, \bar{c} \ldots \bar{n} \cdot\!|\!\cdot \ldots \cdot\!|\!\cdot \bar{a} \ldots \bar{m}, n] \\ &\quad = [a][\bar{b}][\bar{c}] \ldots [\bar{n}] + [\bar{a}][b] \ldots [\bar{n}] \ldots + [\bar{a}] \ldots [\bar{m}][n]. \qquad (3)\end{aligned}$$

For the probability of the occurrence of some one or more of n independent non-exclusive events, we obtain by transposition from the second form of the general addition formula,—

$$\begin{aligned}&[a \cdot\!|\!\cdot b \cdot\!|\!\cdot c \cdot\!|\!\cdot \ldots \cdot\!|\!\cdot n] = [a] + [b] + \ldots + [n] \\ &\quad - [a][b][\bar{c}] \ldots [\bar{n}] - \ldots - [\bar{a}] \ldots [m][n] \\ &\quad - 2[a][b][c][\bar{d}] \ldots [\bar{n}] - \ldots - 2[\bar{a}] \ldots [l][m][n] \\ &\quad - \ldots \\ &\quad - (n-1)[a][b][c] \ldots [n] \qquad (4)\end{aligned}$$

Since the probability of a compound event is the product of the probabilities of the components (when independent), we have the following equation:—

$$[a, b, c \ldots n] = [a][b][c] \ldots [n] \qquad (5)$$

which gives us

$$[a][b][c] \ldots [g] = \frac{[a, b, c \ldots n]}{[h] \ldots [n]}$$

or

$$[a, b \ldots g] = \frac{[a, b, c \ldots n]}{[h, i \ldots n]};$$

that is, the probability of any event is equal to the probability of any compound event into which it enters, divided by the probability of the compound event made up of the remaining components.

We may obtain an expression for the probability of a compound event when the components are not independent, by noticing that in establishing the formula for multiplication the independence of the relations enabled us to substitute in the left-hand member of the equation, $\frac{\rho'}{\infty}$ for $\frac{\rho'\rho''}{\rho''}$. If the relations are not independent, this is not permissible; whence indicating $\frac{\rho'\rho''}{\rho''}$ by $[\rho'\rho'']_{\rho''}$ the equation reads

$$[\rho'\rho'']_{\rho''}\,[\rho''] = [\rho'\rho''],$$

or for identical relations

$$[a, b]_b\,[b] = [a, b],$$

in which $[a, b]_b$ denotes the proportion of a, b's among b's, the probability that an event of the genus b will also be of the species a. An extension of these considerations gives the general formula

$$[a, b \ldots n]_{b \ldots n}\,[b, c . n]_{c \ldots n}\,[c, d . n]_{d \ldots n} \cdots [m, n]_n\,[n] = [a, b \ldots n]; \quad (6)$$

that is, the probability of a compound event, when the components are not independent, is equal to the general probability of any one of the components multiplied by the probability that one of the other components will happen when the first happens, and so on until all the components are exhausted.

Let us suppose that the compound event, instead of being composed of n different events, is composed of n like events, a. If these different occurrences of a are independent, — that is, if the fact that a has occurred once, makes it neither more nor less likely that it will occur again, — we have

$$[a^n] = [a]^n \qquad (7)$$

While the mere fact that a has occurred will not, contrary to the popular notion, make it any more or less likely to recur, it is evident that in many instances attendant circumstances, as in the case of habit, may destroy the independence of successive occurrences.

If a is a compound of independent relations of identity, as $a, b, c, \ldots m$, the formula becomes

$$\begin{aligned} [(a,b,c \ldots m)^n] &= [a,b,c \ldots m]^n \\ &= ([a][b][c] \ldots [m])^n \\ &= [a]^n[b]^n[c]^n \ldots [m]^n; \qquad (8) \end{aligned}$$

that is, the probability of the repetition of a compound event n times is equal to the product of the nth powers of the probabilities of its components.

We have seen that a polynomial relative number expresses the probability of the occurrence of some one or more of the separate events symbolized therein. If the events are exclusives, it expresses the probability of the occurrence of some one of them.

Considering two exclusives, a and b, in order to obtain the probability that one or other of them should occur n times, it is to be noticed first that this event itself is not a single compound event, but a compound alternating event, consisting of as many compound alternatives as there are different arrangements of a and b in n occurrences. Since the probability of an alternating event is the sum of the probabilities of the alternatives, the probability we seek will be the sum of the probabilities of all the compound alternatives; that is, the sum of all the products obtained by forming all possible arrangements of n simple probabilities, each of which must be either $[a]$ or $[b]$. In other words, the operation of finding the probability of the occurrence of one or other of two exclusives n times, is the same as that of raising the binomial $[a] + [b]$ to the nth power. This is otherwise seen thus: Since a and b are exclusives,—

$$[a \mathbin{\cdot|\cdot} b] = [a] + [b];$$

but

$$[(a \mathbin{\cdot|\cdot} b)^n] = [a \mathbin{\cdot|\cdot} b]^n = ([a] + [b])^n.$$

Similarly, for more than two exclusives, the probability of one or other happening p times is equal to the sum of the probabilities of the exclusives raised to the pth power, or

$$[(a \mathbin{\cdot|\cdot} b \mathbin{\cdot|\cdot} c \mathbin{\cdot|\cdot} \ldots \mathbin{\cdot|\cdot} n)^p] = ([a] + [b] + [c] \ldots + [n])^p. \quad (9)$$

It may be observed in relation to the probabilities of the compound alternatives of which these sums are made up, that any one will be equal to all the others in which the elementary exclusives enter in the same proportions, although in different orders. The case of highest probability will evidently be that consisting entirely of that one of the elementary exclusives which has the highest probability, and the case of lowest probability will be that in which the elementary exclusive having the lowest probability alone appears. On the contrary, other considerations show that the most probable proportions in which different alternatives will enter into a series of trials will be the ratios of their probabilities, while the most improbable proportions will be those exhibited by series consisting entirely of some one of the alternatives. The same thing is true of exclusives; the most probable proportion in which they will be found in a series of trials being the ratios of their probabilities. But while with alternatives the sum of the probabilities of all possible orders will continue to be unity, however the number of trials is increased, with exclusives the sum of these probabilities will decrease in geometrical progression as the trials are repeated.

The results thus far reached, readily lead to other combinations of probabilities, as in the following examples: The probability of the occurrence of at least one of two events with a third is given by the equation

$$[(a \cdot|\cdot b), c] = [(a \cdot|\cdot b)][c] = ([a]+[b])[c]-[a][b][c] \quad (10)$$

in which, as in general in probabilities, the events are supposed to be independent.

When a and b are exclusives, the same probability is equal to

$$([a]+[b])\,[c].$$

For any number of exclusives, and any number of other events, the equation becomes

$$[(\alpha \cdot|\cdot \beta \cdot|\cdot \ldots \cdot|\cdot \nu), a, b, \ldots n] =$$
$$([\alpha]+[\beta]+\ldots+[\nu])\,[a][b]\ldots[n] \quad (11)$$

For the probability of the occurrence of one, and only one, of any number of non-exclusive events with any number of others, we have

$$[(\alpha, \bar{\beta} \ldots \bar{\nu} \cdot|\cdot \ldots \cdot|\cdot \ldots \cdot|\cdot \bar{\alpha}, \ldots \bar{\mu}, \nu)\, a, b \ldots n] =$$
$$[a][b]\ldots[n]([\alpha][\bar{\beta}]\ldots[\bar{\nu}]+\ldots+[\bar{\alpha}]\ldots[\bar{\mu}][\nu]) \quad (12)$$

The probability that a will occur m times to n occurrences of b, — that is, that m a's will happen while n b's are happening, — will be the probability of the compound event consisting of m a's and n b's. The probability that m a's will be succeeded by n b's is $[a]^m[b]^n$, and the number of different arrangements of $m+n$ objects, m of one kind and n of another, is $\frac{|\underline{m+n}}{|\underline{m}\;|\underline{n}}$; whence the total probability is

$$\frac{|\underline{m+n}}{|\underline{m}\;|\underline{n}}\,[a]^m\,[b]^n. \quad (13)$$

If a and b were alternating events, this expression would give the probability of the occurrence of some one or other of π exclusives m times, while some one or other of p exclusives is happening n times. Substituting the values assumed in this case by $[a]^m$ and $[b]^n$, we have for this probability

$$\frac{\underline{|m+n}}{\underline{|m}\,\underline{|n}}\left([\alpha]+[\beta]+\ldots+[\pi]\right)^m\left([a]+[b]\ldots+[p]\right)^n. \quad (14)$$

In this investigation of some modes of combining probabilities, suggested by the consideration of Relative Number, we have used the Addition formulæ in reaching (1) the probability of negative events, (2) of some one of n exclusives, (3) of some one, and only one, of n non-exclusives, and (4) of at least one of n non-exclusives. From the Multiplication formulæ we have obtained the probability of a compound event when the components are either (5) independent, or (6) dependent; and by a reference to the involution of Relative Number have established formulæ for the probability of the repetition of (7) simple (8) compound or (9) alternating events. These results have been combined in the more complicated cases (10 – 14) last considered.

A THEORY OF PROBABLE INFERENCE.

By C. S. Peirce.

I.

The following is an example of the simplest kind of probable inference: —

About two per cent of persons wounded in the liver recover;
This man has been wounded in the liver;
Therefore, there are two chances out of a hundred that he will recover.

Compare this with the simplest of syllogisms, say the following: —

Every man dies;
Enoch was a man;
Hence, Enoch must have died.

The latter argument consists in the application of a general rule to a particular case. The former applies to a particular case a rule not absolutely universal, but subject to a known proportion of exceptions. Both may alike be termed deductions, because they bring information about the uniform or usual course of things to bear upon the solution of special questions; and the probable argument may approximate indefinitely to demonstration as the ratio named in the first premise approaches to unity or to zero.

Let us set forth the general formulæ of the two kinds of inference in the manner of formal logic.

FORM I.

Singular Syllogism in Barbara.

Every M is a P;
S is an M;
Hence, S is a P.

FORM II.

Simple Probable Deduction.

The proportion ρ of the M's are P's;
S is an M;
It follows, with probability ρ, that S is a P.

It is to be observed that the ratio ρ need not be exactly specified. We may reason from the premise that not more than two per cent of persons wounded in the liver recover, or from "not less than a certain proportion of the M's are P's," or from "no very large nor very small proportion, etc." In short, ρ is subject to every kind of indeterminacy; it simply excludes some ratios and admits the possibility of the rest.

The analogy between syllogism and what is here called probable deduction is certainly genuine and important; yet how wide the differences between the two modes of inference are, will appear from the following considerations: —

1. The logic of probability is related to ordinary syllogistic as the quantitative to the qualitative branch of the same science. Necessary syllogism recognizes only the inclusion or non-inclusion of one class under another; but probable inference takes account of the proportion

of one class which is contained under a second. It is like the distinction between projective geometry, which asks whether points coincide or not, and metric geometry, which determines their distances.

2. For the existence of ordinary syllogism, all that is requisite is that we should be able to say, in some sense, that one term is contained in another, or that one object stands to a second in one of those relations: "better than," "equivalent to," etc., which are termed *transitive* because if A is in any such relation to B, and B is in the same relation to C, then A is in that relation to C. The universe might be all so fluid and variable that nothing should preserve its individual identity, and that no measurement should be conceivable; and still one portion might remain inclosed within a second, itself inclosed within a third, so that a syllogism would be possible. But probable inference could not be made in such a universe, because no signification would attach to the words "quantitative ratio." For that there must be counting; and consequently units must exist, preserving their identity and variously grouped together.

3. A cardinal distinction between the two kinds of inference is, that in demonstrative reasoning the conclusion follows from the existence of the objective facts laid down in the premises; while in probable reasoning these facts in themselves do not even render the conclusion probable, but account has to be taken of various subjective circumstances,—of the manner in which the premises have been obtained, of there being no countervailing considerations, etc.; in short, good faith and honesty are essential to good logic in probable reasoning.

When the partial rule that the proposition ρ of the M's are P's is applied to show with probability ρ that S is a P, it is requisite, not merely that S should *be* an

M, but also that it should be an instance drawn *at random* from among the *M*'s. Thus, there being four aces in a picquet pack of thirty-two cards, the chance is one eighth that a given card not looked at is an ace; but this is only on the supposition that the card has been drawn at random from the whole pack. If, for instance, it had been drawn from the cards discarded by the players at piquet or euchre, the probability would be quite different. The instance must be drawn at random. Here is a maxim of conduct. The volition of the reasoner (using what machinery it may) has to choose *S* so that it shall be an *M*; but he ought to restrain himself from all further preference, and not allow his will to act in any way that might tend to settle what particular *M* is taken, but should leave that to the operation of chance. Willing and wishing, like other operations of the mind, are *general* and imperfectly determinate. I wish for a horse, — for some particular kind of horse perhaps, but not usually for any individual one. I will to act in a way of which I have a general conception; but so long as my action conforms to that general description, how it is further determined I do not care. Now in choosing the instance *S*, the general intention (including the whole plan of action) should be to select an *M*, but beyond that there should be no preference; and the act of choice should be such that if it were repeated many enough times with the same intention, the result would be that among the totality of selections the different sorts of *M*'s would occur with the same relative frequencies as in experiences in which volition does not intermeddle at all. In cases in which it is found difficult thus to restrain the will by a direct effort, the apparatus of games of chance, — a lottery-wheel, a roulette, cards, or dice, — may be called to our aid. Usually, however, in making a simple probable deduction, we take that instance in which we happen at the time to be interested. In such a case, it is our interest that fulfils the function of an apparatus for random selection; and no better need be desired, so long as we have reason to deem the premise "the proportion ρ of the *M*'s are *P*'s" to be equally true in regard to that part of the *M*'s which are alone likely ever to excite our interest.

Nor is it a matter of indifference in what manner the other premise has been obtained. A card being drawn at random from a picquet pack, the chance is one-eighth that it is an ace, if we have no other knowledge of it. But after we have looked at the card, we can no longer reason in that way. That the conclusion must be drawn in advance of any other knowledge on the subject is a rule that, however elementary, will be found in the sequel to have great importance.

4. The conclusions of the two modes of inference likewise differ. One is necessary; the other only probable. Locke, in the "Essay concerning Human Understanding," hints at the correct analysis of the nature of probability. After remarking that the mathematician positively knows that the sum of the three angles of a triangle is equal to two right angles because he apprehends the geometrical proof, he then continues: "But another man who never took the pains to observe the demonstration, hearing a mathematician, a man of credit, affirm the three angles of a triangle to be equal to two right ones, *assents* to it; that is, receives it for true. In which case, the foundation of his assent is the probability of the thing, the proof being such as, for the most part, carries truth with it; the man on whose testimony he receives it not being wont to affirm anything contrary to or besides his knowledge,

especially in matters of this kind." Those who know Locke are accustomed to look for more meaning in his words than appears at first glance. There is an allusion in this passage to the fact that a probable argument is always regarded as belonging to a *genus* of arguments. This is, in fact, true of any kind of argument. For the belief expressed by the conclusion is determined or caused by the belief expressed by the premises. There is, therefore, some general rule according to which the one succeeds the other. But, further, the reasoner is conscious of there being such a rule, for otherwise he would not know he was reasoning, and could exercise no attention or control; and to such an involuntary operation the name reasoning is very properly not applied. In all cases, then, we are conscious that our inference belongs to a general class of logical forms, although we are not necessarily able to describe the general class. The difference between necessary and probable reasoning is that in the one case we conceive that such facts as are expressed by the premises are never, in the whole range of possibility, true, without another fact, related to them as our conclusion is to our premises, being true likewise; while in the other case we merely conceive that, in reasoning as we do, we are following a general maxim that will usually lead us to the truth.

So long as there are exceptions to the rule that all men wounded in the liver die, it does not necessarily follow that because a given man is wounded in the liver he cannot recover. Still, we know that if we were to reason in that way, we should be following a mode of inference which would only lead us wrong, in the long run, once in fifty times; and this is what we mean when we say that the probability is one out of fifty that the man will recover. To say, then, that a proposition has the probability ρ means that to infer it to be true would be to follow an argument such as would carry truth with it in the ratio of frequency ρ.

It is plainly useful that we should have a stronger feeling of confidence about a sort of inference which will oftener lead us to the truth than about an inference that will less often prove right, — and such a sensation we do have. The celebrated law of Fechner is, that as the force acting upon an organ of sense increases in geometrical progression, the intensity of the sensation increases in arithmetical progression. In this case the odds (that is, the ratio of the chances in favor of a conclusion to the chances against it) take the place of the exciting cause, while the sensation itself is the feeling of confidence. When two arguments tend to the same conclusion, our confidence in the latter is equal to the sum of what the two arguments separately would produce; the *odds* are the product of the *odds* in favor of the two arguments separately. When the value of the *odds* reduces to unity, our confidence is null; when the *odds* are less than unity, we have more or less confidence in the negative of the conclusion.

II.

The principle of probable deduction still applies when S, instead of being a single M, is a set of M's, — n in number. The reasoning then takes the following form: —

FORM III.

Complex Probable Deduction.

Among all sets of n M's, the proportion q consist each of m P's and of $n - m$ not-P's;

S, *S*′, *S*″, etc. form a set of *n* objects drawn at random from among the *M*'s:

Hence, the probability is *q* that among *S*, *S*′, *S*″, etc. there are *m* *P*'s and $n-m$ not-*P*'s.

In saying that *S*, *S*′, *S*″, etc. form a set drawn at random, we here mean that not only are the different individuals drawn at random, but also that they are so drawn that the qualities which may belong to one have no influence upon the selection of any other. In other words, the individual drawings are independent, and the set as a whole is taken at random from among all possible sets of *n* *M*'s. In strictness, this supposes that the same individual may be drawn several times in the same set, although if the number of *M*'s is large compared with *n*, it makes no appreciable difference whether this is the case or not.

The following formula expresses the proportion, among all sets of *n* *M*'s, of those which consist of *m* *P*'s and $n-m$ not-*P*'s. The letter *r* denotes the proportion of *P*'s among the *M*'s, and the sign of admiration is used to express the continued product of all integer numbers from 1 to the number after which it is placed. Thus, $4! = 1 . 2 . 3 . 4 = 24$, etc. The formula is

$$q = n! \times \frac{r^m}{m!} \times \frac{(1-r)^{n-m}}{(n-m)!}$$

As an example, let us assume the proportion $r = \frac{2}{3}$ and the number of *M*'s in a set $n = 15$. Then the values of the probability *q* for different numbers, *m*, of *P*'s, are fractions having for their common denominator 14,348,907, and for their numerators as follows: —

m	Numerator of q.	m	Numerator of q.
0	1	8	1667360
1	30	9	2562560
2	420	10	3075072
3	3640	11	2795520
4	21840	12	1863680
5	96096	13	860160
6	320320	14	122880
7	823680	15	32768

A very little mathematics would suffice to show that, r and n being fixed, q always reaches its maximum value with that value of m that is next less than $(n+1)r$,* and that q is very small unless m has nearly this value.

Upon these facts is based another form of inference to which I give the name of statistical deduction. Its general formula is as follows: —

FORM IV.

Statistical Deduction.

The proportion r of the M's are P's;

S', S'', S''', etc., are a *numerous* set, taken at random from among the M's:

Hence, *probably and approximately*, the proportion r of the S's are P's.

As an example, take this: —

A little more than half of all human births are males;

Hence, probably a little over half of all the births in New York during any one year are males.

We have now no longer to deal with a mere probable inference, but with a *probable approximate* inference.

* In case $(n+1)r$ is a whole number, q has equal values for $m = (n+1)r$ and for $m = (n+1)r - 1$.

This conception is a somewhat complicated one, meaning that the probability is greater according as the limits of approximation are wider, conformably to the mathematical expression for the values of q.

This conclusion has no meaning at all unless there be more than one instance; and it has hardly any meaning unless the instances are somewhat numerous. When this is the case, there is a more convenient way of obtaining (not exactly, but quite near enough for all practical purposes) either a single value of q or the sum of successive values from $m = m_1$ to $m = m_2$ inclusive. The rule is first to calculate two quantities which may conveniently be called t_1 and t_2 according to these formulæ: —

$$t_1 = \frac{m_1 - (n+1)r}{\sqrt{2nr(1-r)}} \qquad t_2 = \frac{1 + m_2 - (n+1)r}{\sqrt{2nr(1-r)}}$$

where $m_2 > m_1$. Either or both the quantities t_1 and t_2 may be negative. Next with each of these quantities enter the table below, and take out $\frac{1}{2}\,\Theta t_1$ and $\frac{1}{2}\,\Theta t_2$ and give each the same sign as the t from which it is derived. Then

$$\Sigma\, q = \tfrac{1}{2}\,\Theta t_2 - \tfrac{1}{2}\,\Theta t_1.$$

$$\textit{Table of } \Theta t = \frac{2}{\sqrt{\pi}}\int_0^t e^{-t^2}\,dt.$$

t	Θt	t	Θt	t	Θt
0.0	0.000	1.0	$0.84\bar{3}$	2.0	0.99532
0.1	0.112	1.1	0.880	2.1	0.99702
0.2	$0.22\bar{3}$	1.2	0.910	2.2	$0.9981\bar{4}$
0.3	$0.32\bar{9}$	1.3	0.934	2.3	$0.9988\bar{6}$
0.4	0.428	1.4	0.952	2.4	0.99931
0.5	0.520	1.5	0.966	2.5	0.99959
0.6	$0.60\bar{4}$	1.6	0.976	2.6	0.99976
0.7	$0.67\bar{8}$	1.7	$0.98\bar{4}$	2.7	$0.9998\bar{7}$
0.8	0.742	1.8	0.989	2.8	0.99992
0.9	$0.79\bar{7}$	1.9	$0.99\bar{3}$	2.9	0.99996
1.0	$0.84\bar{3}$	2.0	0.995	3.0	$0.9999\bar{8}$

t	Θ
4	0.999999989
5	0.9999999999984
6	0.999999999999999982
7	0.999999999999999999999958

In rough calculations we may take Θt equal to t for t less than 0.7, and as equal to *unity* for any value above $t = 1.4$.

The principle of statistical deduction is that these two proportions, — namely, that of the P's among the M's, and that of the P's among the S's, — are probably and approximately equal. If, then, this principle justifies our inferring the value of the second proportion from the known value of the first, it equally justifies our inferring the value of the first from that of the second, if the first is unknown but the second has been observed. We thus obtain the following form of inference:—

FORM V.

Induction.

S', S'', S''', etc., form a numerous set taken at random from among the M's;

S', S'', S''', etc., are found to be — the proportion ρ of them — P's:

Hence, *probably* and *approximately* the same proportion, ρ, of the M's are P's.

The following are examples. From a bag of coffee a handful is taken out, and found to have nine tenths of the beans perfect; whence it is inferred that about nine-tenths of all the beans in the bag are probably perfect. The United States Census of 1870 shows that of native white children under one year old, there were 478,774 males to 463,320 females; while of colored children of the same age there were 75,985 males to 76,637 females. We infer that generally there is a larger proportion of female births among negroes than among whites.

When the ratio ρ is *unity* or *zero*, the inference is an ordinary induction; and I ask leave to extend the term induction to all such inference, whatever be the value of ρ. It is, in fact, inferring from a sample to the whole lot sampled. These two forms of inference, statistical deduction and induction, plainly depend upon the same principle of equality of ratios, so that their validity is the same. Yet the nature of the probability in the two cases is very different. In the statistical deduction, we know that among the whole body of M's the proportion of P's is ρ; we say, then, that the S's being random drawings of M's are probably P's in about the same proportion, — and though this may happen not to be so, yet at any rate, on continuing the drawing sufficiently, our prediction of the ratio will be vindicated at last. On the other hand, in induction we say that the proportion ρ of the sample being P's, probably there is about the same proportion in the whole lot; or at least, if this happens not to be so, then on continuing the drawings the inference will be, not *vindicated* as in the other case, but *modified* so as to become true. The deduction, then, is probable in this sense, that though its conclusion may in a particular case be falsified, yet similar conclusions (with the same ratio ρ) would generally prove approximately true; while the induction is probable in this sense, that though it may happen to give a false conclusion, yet in most cases in which the same precept of inference was followed, a different and approximately true inference (with the right value of ρ) would be drawn.

IV.

Before going any further with the study of Form V., I wish to join to it another extremely analogous form.

We often speak of one thing being very much like another, and thus apply a vague quantity to resemblance. Even if qualities are not subject to exact numeration, we may conceive them to be approximately measurable. We may then measure resemblance by a scale of numbers from zero up to unity. To say that S has a 1-likeness to a P will mean that it has every character of a P, and consequently *is* a P. To say that it has a 0-likeness will imply total dissimilarity. We shall then be able to reason as follows: —

FORM II. (*bis*).

Simple probable deduction in depth.

Every M has the simple mark P;
The S's have an r-likeness to the M's:
Hence, the probability is r that every S is P.

It would be difficult, perhaps impossible, to adduce an example of such kind of inference, for the reason that *simple marks* are not known to us. We may, however, illustrate the complex probable deduction in depth (the general form of which it is not worth while to set down) as follows: I forget whether, in the ritualistic churches, a bell is tinkled at the elevation of the Host or not. Knowing, however, that the services resemble somewhat decidedly those of the Roman Mass, I think that it is not unlikely that the bell is used in the ritualistic, as in the Roman, churches.

We shall also have the following:—

FORM IV. (*bis*).

Statistical deduction in depth.

Every M has, for example, the numerous marks P', P'', P''', etc.
S has an r-likeness to the M's:
Hence, probably and approximately, S has the proportion r of the marks P', P'', P''', etc.

For example, we know that the French and Italians are a good deal alike in their ideas, characters, temperaments, genius, customs, institutions, etc., while they also differ very markedly in all these respects. Suppose, then, that I know a boy who is going to make a short trip through France and Italy; I can safely predict that among the really numerous though relatively few respects in which he will be able to compare the two people, about the same degree of resemblance will be found.

Both these modes of inference are clearly deductive. When $r = 1$, they reduce to Barbara.[1]

Corresponding to induction, we have the following mode of inference:—

FORM V. (*bis*).

Hypothesis.

M has, for example, the numerous marks P', P'', P''', etc.
S has the proportion r of the marks P', P'', P''', etc.:
Hence, probably and approximately, *S* has an r-likeness to *M*.

Thus, we know, that the ancient Mound-builders of North America present, in all those respects in which we have been able to make the comparison, a limited degree of resemblance with the Pueblo Indians. The inference is, then, that in all respects there is about the same degree of resemblance between these races.

If I am permitted the extended sense which I have given to the word "induction," this argument is simply an induction respecting qualities instead of respecting

[1] When $r = 0$, the last form becomes

M has all the marks *P*;
S has no mark of *M*:
Hence, *S* has none of the marks *P*.

When the universe of marks is unlimited (see a note appended to this paper for an explanation of this expression), the only way in which two terms can fail to have a common mark is by their together filling the universe of things; and consequently this form then becomes,

M is *P*;
Every non-*S* is *M*:
Hence, every non-*S* is *P*.

This is one of De Morgan's syllogisms.

In putting $r = 0$ in Form II. (*bis*) it must be noted that, since *P* is simple in depth, to say that *S* is not *P* is to say that it has no mark of *P*.

things. In point of fact P', P'', P''', etc. constitute a random sample of the characters of M, and the ratio r of them being found to belong to S, the same ratio of all the characters of M are concluded to belong to S. This kind of argument, however, as it actually occurs, differs very much from induction, owing to the impossibility of simply counting qualities as individual things are counted. Characters have to be weighed rather than counted. Thus, antimony is bluish-gray: that is a character. Bismuth is a sort of rose-gray; it is decidedly different from antimony in color, and yet not so very different as gold, silver, copper, and tin are.

I call this induction of characters *hypothetic inference*, or, briefly, *hypothesis*. This is perhaps not a very happy designation, yet it is difficult to find a better. The term "hypothesis" has many well established and distinct meanings. Among these is that of a proposition believed in because its consequences agree with experience. This is the sense in which Newton used the word when he said, *Hypotheses non fingo.* He meant that he was merely giving a general formula for the motions of the heavenly bodies, but was not undertaking to mount to the causes of the acceleration they exhibit. The inferences of Kepler, on the other hand, were hypotheses in this sense; for he traced out the miscellaneous consequences of the supposition that Mars moved in an ellipse, with the sun at the focus, and showed that both the longitudes and the latitudes resulting from this theory were such as agreed with observation. These two components of the motion were observed; the third, that of approach to or regression from the earth, was supposed. Now, if in Form V. (*bis*) we put $r = 1$, the inference is the drawing of a hypothesis in this sense. I take the liberty of extending the use of the word by permitting r to have any value from zero to unity. The term is certainly not all that could be desired; for the word hypothesis, as ordinarily used, carries with it a suggestion of uncertainty, and of something to be superseded, which does not belong at all to my use of it. But we must use existing language as best we may, balancing the reasons for and against any mode of expression, for none is perfect; at least the term is not so utterly misleading as "analogy" would be, and with proper explanation it will, I hope, be understood.

V.

The following examples will illustrate the distinction between statistical deduction, induction, and hypothesis. If I wished to order a font of type expressly for the printing of this book, knowing, as I do, that in all English writing the letter *e* occurs oftener than any other letter, I should want more *e's* in my font than other letters. For what is true of all other English writing is no doubt true of these papers. This is a statistical deduction. But then the words used in logical writings are rather peculiar, and a good deal of use is made of single letters. I might, then, count the number of occurrences of the different letters upon a dozen or so pages of the manuscript, and thence conclude the relative amounts of the different kinds of type required in the font. That would be inductive inference. If now I were to order the font, and if, after some days, I were to receive a box containing a large number of little paper parcels of very different sizes, I should naturally infer that this was the font of types I had ordered; and this would be hypothetic inference. Again, if a dispatch in cipher is captured, and it is found to be written with twenty-six characters, one of which occurs much more frequently than any of the others, we are at once led to suppose that each character represents a letter, and that the one occurring so frequently stands for *e*. This is also hypothetic inference.

We are thus led to divide all probable reasoning into deductive and ampliative, and further to divide ampliative reasoning into induction and hypothesis. In deductive reasoning, though the predicted ratio may be wrong in a limited number of drawings, yet it will be approximately verified in a larger number. In ampliative reasoning the ratio may be wrong, because the inference is based on but a limited number of instances; but on enlarging the sample the ratio will be changed till it becomes approximately correct. In induction, the instances drawn at random are numerable things; in hypothesis they are characters, which are not capable of strict enumeration, but have to be otherwise estimated.

This classification of probable inference is connected with a preference for the copula of inclusion over those used by Miss Ladd and by Mr. Mitchell.[1] De Morgan established eight forms of simple propositions; and from a purely formal point of view no one of these has a right to be considered as more fundamental than any other. But formal logic must not be too purely formal; it must represent a fact of psychology, or else it is in danger of degenerating into a mathematical recreation. The categorical proposition, "every man is mortal," is but a modification of the hypothetical proposition, "if humanity, then mortality;" and since the very first conception from which logic springs is that one proposition follows from another, I hold that "if *A*, then *B*" should be taken as the typical form of judgment. Time flows; and, in time, from one state of belief (represented by the premises of an argu-

[1] I do not here speak of Mr. Jevons, because my objection to the copula of identity is of a somewhat different kind.

ment) another (represented by its conclusion) is developed. Logic arises from this circumstance, without which we could not learn anything nor correct any opinion. To say that an inference is correct is to say that if the premises are true the conclusion is also true; or that every possible state of things in which the premises should be true would be included among the possible states of things in which the conclusion would be true. We are thus led to the copula of inclusion. But the main characteristic of the relation of inclusion is that it is transitive, — that is, that what is included in something included in anything is itself included in that thing; or, that if *A* is *B* and *B* is *C*, then *A* is *C*. We thus get *Barbara* as the primitive type of inference. Now in Barbara we have a *Rule*, a *Case* under the *Rule*, and the inference of the *Result* of that rule in that case. For example: —

Rule. All men are mortal;
Case. Enoch was a man.
Result. Enoch was mortal.

The cognition of a rule is not necessarily conscious, but is of the nature of a habit, acquired or congenital. The cognition of a case is of the general nature of a sensation; that is to say, it is something which comes up into present consciousness. The cognition of a result is of the nature of a decision to act in a particular way on a given occasion.[1] In point of fact, a syllogism in Barbara virtually takes place when we irritate the foot of a decapitated frog. The connection between the afferent and efferent nerve, whatever it may be, constitutes a nervous habit, a rule of action, which is the physio-

[1] See my paper on "How to make our ideas clear." — *Popular Science Monthly*, January, 1878.

logical analogue of the major premise. The disturbance of the ganglionic equilibrium, owing to the irritation, is the physiological form of that which, psychologically considered, is a sensation; and, logically considered, is the occurrence of a case. The explosion through the efferent nerve is the physiological form of that which psychologically is a volition, and logically the inference of a result. When we pass from the lowest to the highest forms of inervation, the physiological equivalents escape our observation; but, psychologically, we still have, first, habit, — which in its highest form is understanding, and which corresponds to the major premise of *Barbara;* we have, second, feeling, or present consciousness, corresponding to the minor premise of *Barbara;* and we have, third, volition, corresponding to the conclusion of the same mode of syllogism. Although these analogies, like all very broad generalizations, may seem very fanciful at first sight, yet the more the reader reflects upon them the more profoundly true I am confident they will appear. They give a significance to the ancient system of formal logic which no other can at all share.

Deduction proceeds from Rule and Case to Result; it is the formula of Volition. Induction proceeds from Case and Result to Rule; it is the formula of the formation of a habit or general conception, — a process which, psychologically as well as logically, depends on the repetition of instances or sensations. Hypothesis proceeds from Rule and Result to Case; it is the formula of the acquirement of secondary sensation, — a process by which a confused concatenation of predicates is brought into order under a synthetizing predicate.

We usually conceive Nature to be perpetually making deductions in Barbara. This is our natural and anthropomorphic metaphysics. We conceive that there are

Laws of Nature, which are her Rules or major premises. We conceive that Cases arise under these laws; these cases consist in the predication, or occurrence, of *causes*, which are the middle terms of the syllogisms. And, finally, we conceive that the occurrence of these causes, by virtue of the laws of Nature, result in effects which are the conclusions of the syllogisms. Conceiving of nature in this way, we naturally conceive of science as having three tasks, — (1) the discovery of Laws, which is accomplished by induction; (2) the discovery of Causes, which is accomplished by hypothetic inference; and (3) the prediction of Effects, which is accomplished by deduction. It appears to me to be highly useful to select a system of logic which shall preserve all these natural conceptions.

It may be added that, generally speaking, the conclusions of Hypothetic Inference cannot be arrived at inductively, because their truth is not susceptible of direct observation in single cases. Nor can the conclusions of Inductions, on account of their generality, be reached by hypothetic inference. For instance, any historical fact, as that Napoleon Bonaparte once lived, is a hypothesis; we believe the fact, because its effects — I mean current tradition, the histories, the monuments, etc. — are observed. But no mere generalization of observed facts could ever teach us that Napoleon lived. So we inductively infer that every particle of matter gravitates toward every other. Hypothesis might lead to this result for any given pair of particles, but it never could show that the law was universal.

VI.

We now come to the consideration of the Rules which have to be followed in order to make valid and strong

Inductions and Hypotheses. These rules can all be reduced to a single one; namely, that the statistical deduction of which the Induction or Hypothesis is the inversion, must be valid and strong.

We have seen that Inductions and Hypotheses are inferences from the conclusion and one premise of a statistical syllogism to the other premise. In the case of hypothesis, this syllogism is called the *explanation.* Thus in one of the examples used above, we suppose the cryptograph to be an English cipher, because, as we say, this *explains* the observed phenomena that there are about two dozen characters, that one occurs more frequently than the rest, especially at the ends of words, etc. The explanation is, —

> Simple English ciphers have certain peculiarities;
> This is a simple English cipher:
> Hence, this necessarily has these peculiarities.

This explanation is present to the mind of the reasoner, too; so much so, that we commonly say that the hypothesis is adopted *for the sake of* the explanation. Of induction we do not, in ordinary language, say that it explains phenomena; still, the statistical deduction, of which it is the inversion, plays, in a general way, the same part as the explanation in hypothesis. From a barrel of apples, that I am thinking of buying, I draw out three or four as a sample. If I find the sample somewhat decayed, I ask myself, in ordinary language, not "Why is this?" but "How is this?" And I answer that it probably comes from nearly all the apples in the barrel being in bad condition. The distinction between the "Why" of hypothesis and the "How" of induction is not very great; both ask for a statistical syllogism, of which the observed fact shall be the conclusion, the known conditions of the observation one premise, and the inductive or hypothetic inference the other. This statistical syllogism may be conveniently termed the explanatory syllogism.

In order that an induction or hypothesis should have any validity at all, it is requisite that the explanatory syllogism should be a valid statistical deduction. Its conclusion must not merely follow from the premises, but follow from them upon the principle of probability. The inversion of *ordinary* syllogism does not give rise to an induction or hypothesis. The statistical syllogism of Form IV. is invertible, because it proceeds upon the principle of an approximate *equality* between the ratio of P's in the whole class and the ratio in a well-drawn sample, and because equality is a convertible relation. But ordinary syllogism is based upon the property of the relation of containing and contained, and that is not a convertible relation. There is, however, a way in which ordinary syllogism may be inverted; namely, the conclusion and either of the premises may be interchanged by negativing each of them. This is the way in which the indirect, or apagogical,[1] figures of syllogism are derived from the first, and in which the *modus tollens* is derived from the *modus ponens*. The following schemes show this: —

First Figure.

Rule.	All M is P;
Case.	S is M:
Result.	S is P.

Second Figure.

Rule.	All M is P;
Denial of Result.	S is not P:
Denial of Case.	S is not M.

Third Figure.

Denial of Result.	S is not P;
Case.	S is M:
Denial of Rule.	Some M is not P.

[1] From *apagoge*, Aristotle's name for the *reductio ad absurdum*.

Modus Ponens.

Rule. If A is true, C is true;
Case. In a certain case A is true:
Result. ∴ In that case C is true.

Modus Tollens.

Rule. If A is true, C is true;
Denial of Result. In a certain case C is not true:
Denial of Case. ∴ In that case A is not true.

Modus Innominatus.

Case. In a certain case A is true;
Denial of Result. In that case C is not true:
Denial of Rule. ∴ If A is true, C is not necessarily true.

Now suppose we ask ourselves what would be the result of thus apagogically inverting a statistical deduction. Let us take, for example, Form IV: —

The S's are a numerous random sample of the M's;
The proportion r of the M's are P's:
Hence, probably about the proportion r of the S's are P's.

The ratio r, as we have already noticed, is not necessarily perfectly definite; it may be only known to have a certain maximum or minimum; in fact, it may have any kind of indeterminacy. Of all possible values between 0 and 1, it admits of some and excludes others. The logical negative of the ratio r is, therefore, itself a ratio, which we may name ρ; it admits of every value which r excludes, and excludes every value of which r admits. Transposing, then, the major premise and conclusion of our statistical deduction, and at the same time denying both, we obtain the following inverted form: —

The *S*'s are a numerous random sample of the *M*'s;
The proportion ρ of the *S*'s are *P*'s:
Hence, probably about the proportion ρ of the *M*'s are *P*'s.[1]

But this coincides with the formula of Induction. Again, let us apagogically invert the statistical deduction of Form IV. (*bis*). This form is, —

Every *M* has, for example, the numerous marks P', P'', P''', etc.
S has an *r*-likeness to the *M*'s:
Hence, probably and approximately, *S* has the proportion *r* of the marks P', P'', P''', etc.

Transposing the minor premise and conclusion, at the same time denying both, we get the inverted form, —

Every *M* has, for example, the numerous marks P', P'', P''', etc.
S has the proportion ρ of the marks P', P'', P''', etc.:
Hence, probably and approximately, *S* has a ρ-likeness to the class of *M*'s.

This coincides with the formula of Hypothesis. Thus we see that Induction and Hypothesis are nothing but the apagogical inversions of statistical deductions. Accordingly, when *r* is taken as 1, so that ρ is "less than 1," or when *r* is taken as 0, so that ρ is "more than 0," the induction degenerates into a syllogism of the third figure and the hypothesis into a syllogism of the second figure.

[1] The conclusion of the statistical deduction is here regarded as being "the proportion *r* of the *S*'s are *P*'s," and the words "probably about" as indicating the modality with which this conclusion is drawn and held for true. It would be equally true to consider the "probably about" as forming part of the contents of the conclusion; only from that point of view the inference ceases to be probable, and becomes rigidly necessary, and its apagogical inversion is also a necessary inference presenting no particular interest.

In these special cases, there is no very essential difference between the mode of reasoning in the direct and in the apagogical form. But, in general, while the probability of the two forms is precisely the same, — in this sense, that for any fixed proportion of *P*'s among the *M*'s (or of marks of *S*'s among the marks of the *M*'s) the probability of any given error in the concluded value is precisely the same in the indirect as it is in the direct form, — yet there is this striking difference, that a multiplication of instances will in the one case *confirm*, and in the other *modify*, the concluded value of the ratio.

We are thus led to another form for our rule of validity of ampliative inference; namely, instead of saying that the *explanatory* syllogism must be a good probable deduction, we may say that the syllogism of which the induction or hypothesis is the apagogical modification (in the traditional language of logic, the *reduction*) must be valid.

Probable inferences, though valid, may still differ in their strength. A probable deduction has a greater or less probable error in the concluded ratio. When r is a definite number the probable error is also definite; but as a general rule we can only assign maximum and minimum values of the probable error. The probable error is, in fact, —

$$0.477 \sqrt{\frac{2\,r\,(1-r)}{n}}$$

where n is the number of independent instances. The same formula gives the probable error of an induction or hypothesis; only that in these cases, r being wholly indeterminate, the minimum value is *zero*, and the maximum is obtained by putting $r = \frac{1}{2}$.

VII.

Although the rule given above really contains all the conditions to which Inductions and Hypotheses need to conform, yet inasmuch as there are many delicate questions in regard to the application of it, and particularly since it is of that nature that a violation of it, if not too gross, may not absolutely destroy the virtue of the reasoning, a somewhat detailed study of its requirements in regard to each of the premises of the argument is still needed.

The first premise of a scientific inference is that certain things (in the case of induction) or certain characters (in the case of hypothesis) constitute a fairly chosen *sample* of the class of things or the run of characters from which they have been drawn.

The rule requires that the sample should be drawn at random and independently from the whole lot sampled. That is to say, the sample must be taken according to a precept or method which, being applied over and over again indefinitely, would in the long run result in the drawing of any one set of instances as often as any other set of the same number.

The needfulness of this rule is obvious; the difficulty is to know how we are to carry it out. The usual method is mentally to run over the lot of objects or characters to be sampled, abstracting our attention from their peculiarities, and arresting ourselves at this one or that one from motives wholly unconnected with those peculiarities. But this abstention from a further determination of our choice often demands an effort of the will that is beyond our strength; and in that case a mechanical contrivance may be called to our aid. We may, for example, number all the objects of the lot, and then draw numbers by means of a roulette, or other such instrument. We may even go so far as to say that this method is the type of all random drawing; for when we abstract our attention from the peculiarities of objects, the psychologists tell us that what we do is to substitute for the images of sense certain mental signs, and when we proceed to a random and arbitrary choice among these abstract objects we are governed by fortuitous determinations of the nervous system, which in this case serves the purpose of a roulette.

The drawing of objects at random is an act in which honesty is called for; and it is often hard enough to be sure that we have dealt honestly with ourselves in the matter, and still more hard to be satisfied of the honesty of another. Accordingly, one method of sampling has come to be preferred in argumentation; namely, to take of the class to be sampled all the objects of which we have a sufficient knowledge. Sampling is, however, a real art, well deserving an extended study by itself: to enlarge upon it here would lead us aside from our main purpose.

Let us rather ask what will be the effect upon inductive inference of an imperfection in the strictly random character of the sampling. Suppose that, instead of using such a precept of selection that any one M would in the long run be chosen as often as any other, we used a precept which would give a preference to a certain half of the M's, so that they would be drawn twice as often as the rest. If we were to draw a numerous sample by such a precept, and if we were to find that the proportion ρ of the sample consisted of P's, the inference that we should be regularly entitled to make would be, that among all the M's, counting the preferred half for two each, the proportion ρ would be P's. But this regular inductive inference being granted, from it we could deduce by arithmetic the further conclusion that, counting the *M*'s for one each, the proportion of *P*'s among them must (ρ being over $\frac{2}{3}$) lie between $\frac{3}{4}\rho + \frac{1}{4}$ and $\frac{3}{2}\rho - \frac{1}{2}$. Hence, if more than two thirds of the instances drawn by the use of the false precept were found to be *P*'s, we should be entitled to conclude that more than half of all the *M*'s were *P*'s. Thus, without allowing ourselves to be led away into a mathematical discussion, we can easily see that, in general, an imperfection of that kind in the random character of the sampling will only weaken the inductive conclusion, and render the concluded ratio less determinate, but will not necessarily destroy the force of the argument completely. In particular, when ρ approximates towards 1 or 0, the effect of the imperfect sampling will be but slight.

Nor must we lose sight of the constant tendency of the inductive process to correct itself. This is of its essence. This is the marvel of it. The probability of its conclusion only consists in the fact that if the true value of the ratio sought has not been reached, an extension of the inductive process will lead to a closer approximation. Thus, even though doubts may be entertained whether one selection of instances is a random one, yet a different selection, made by a different method, will be likely to vary from the normal in a different way, and if the ratios derived from such different selections are nearly equal, they may be presumed to be near the truth. This consideration makes it extremely advantageous in all ampliative reasoning to fortify one method of investigation by another.[1] Still we must not allow ourselves to trust so

[1] This I conceive to be all the truth there is in the doctrine of Bacon and Mill regarding different Methods of Experimental Inquiry. The main proposition of Bacon and Mill's doctrine is, that in order to prove that all *M*'s are *P*'s, we should not only take random instances of the *M*'s and

much to this virtue of induction as to relax our efforts towards making our drawings of instances as random and independent as we can. For if we infer a ratio from a number of different inductions, the magnitude of its probable error will depend very much more on the worst than on the best inductions used.

We have, thus far, supposed that although the selection of instances is not exactly regular, yet the precept followed is such that every unit of the lot would eventually get drawn. But very often it is impracticable so to draw our instances, for the reason that a part of the lot to be sampled is absolutely inaccessible to our powers of observation. If we want to know whether it will be profitable to open a mine, we sample the ore; but in advance of our mining operations, we can obtain only what ore lies near the surface. Then, simple induction becomes worthless, and another method must be resorted to. Suppose we wish to make an induction regarding a series of events extending from the distant past to the distant future; only those events of the series which occur within the period of time over which available history extends can be taken as instances. Within this period we may find that the events of the class in question present some uniform character; yet how do we know but this uniformity was suddenly established a little while before the history commenced, or will suddenly break up a little while after it terminates? Now, whether the uniformity

examine them to see that they are *P*'s, but we should also take instances of not-*P*'s and examine them to see that they are not-*M*'s. This is an excellent way of fortifying one induction by another, when it is applicable; but it is entirely inapplicable when *r* has any other value than 1 or 0. For, in general, there is no connection between the proportion of *M*'s that are *P*'s and the proportion of non-*P*'s that are non-*M*'s. A very small proportion of calves may be monstrosities, and yet a very large proportion of monstrosities may be calves.

observed consists (1) in a mere resemblance between all the phenomena, or (2) in their consisting of a disorderly mixture of two kinds in a certain constant proportion, or (3) in the character of the events being a mathematical function of the time of occurrence, — in any of these cases we can make use of an apagoge from the following probable deduction: —

Within the period of time *M*, a certain event *P* occurs;
S is a period of time taken at random from *M*, and more than half as long:
Hence, probably the event *P* will occur within the time *S*.

Inverting this deduction, we have the following ampliative inference: —

S is a period of time taken at random from *M*, and more than half as long;
The event *P* does not happen in the time *S*:
Hence, probably the event *P* does not happen in the period *M*.

The probability of the conclusion consists in this, that we here follow a precept of inference, which, if it is very often applied, will more than half the time lead us right. Analogous reasoning would obviously apply to any portion of an unidimensional continuum, which might be similar to periods of time. This is a sort of logic which is often applied by physicists in what is called *extrapolation* of an empirical law. As compared with a typical induction, it is obviously an excessively weak kind of inference. Although indispensable in almost every branch of science, it can lead to no solid conclusions in regard to what is remote from the field of direct perception, unless it be bolstered up in certain ways to which we shall have occasion to refer further on.

Let us now consider another class of difficulties in regard to the rule that the samples must be drawn at random and independently. In the first place, what if the lot to be sampled be infinite in number? In what sense could a random sample be taken from a lot like that? A random sample is one taken according to a method that would, in the long run, draw any one object as often as any other. In what sense can such drawing be made from an infinite class? The answer is not far to seek. Conceive a cardboard disk revolving in its own plane about its centre, and pretty accurately balanced, so that when put into rotation it shall be about[1] as likely to come to rest in any one position as in any other; and let a fixed pointer indicate a position on the disk: the number of points on the circumference is infinite, and on rotating the disk repeatedly the pointer enables us to make a selection from this infinite number. This means merely that although the points are innumerable, yet there is a certain order among them that enables us to run them through and pick from them as from a very numerous collection. In such a case, and in no other, can an infinite lot be sampled. But it would be equally true to say that a finite lot can be sampled only on condition that it can be regarded as equivalent to an infinite lot. For the random sampling of a finite class supposes the possibility of drawing out an object, throwing it back, and continuing this process indefinitely; so that what is really sampled is not the finite collection of things, but the unlimited number of possible drawings.

But though there is thus no insuperable difficulty in sampling an infinite lot, yet it must be remembered that the conclusion of inductive reasoning only consists in the

[1] I say *about*, because the doctrine of probability only deals with approximate evaluations.

approximate evaluation of a *ratio*, so that it never can authorize us to conclude that in an infinite lot sampled there exists no single exception to a rule. Although all the planets are found to gravitate toward one another, this affords not the slightest direct reason for denying that among the innumerable orbs of heaven there may be some which exert no such force. Although at no point of space where we have yet been have we found any possibility of motion in a fourth dimension, yet this does not tend to show (by simple induction, at least) that space has absolutely but three dimensions. Although all the bodies we have had the opportunity of examining appear to obey the law of inertia, this does not prove that atoms and atomicules are subject to the same law. Such conclusions must be reached, if at all, in some other way than by simple induction. This latter may show that it is unlikely that, in my lifetime or yours, things so extraordinary should be found, but do not warrant extending the prediction into the indefinite future. And experience shows it is not safe to predict that such and such a fact will *never* be met with.

If the different instances of the lot sampled are to be drawn independently, as the rule requires, then the fact that an instance has been drawn once must not prevent its being drawn again. It is true that if the objects remaining unchosen are very much more numerous than those selected, it makes practically no difference whether they have a chance of being drawn again or not, since that chance is in any case very small. Probability is wholly an affair of approximate, not at all of exact, measurement; so that when the class sampled is very large, there is no need of considering whether objects can be drawn more than once or not. But in what is known as "reasoning from analogy," the class sampled is small, and no instance is taken twice. For example: we know that of the major planets the Earth, Mars, Jupiter, and Saturn revolve on their axes, and we conclude that the remaining four, Mercury, Venus, Uranus, and Neptune, probably do the like. This is essentially different from an inference from what has been found in drawings made hitherto, to what will be found in indefinitely numerous drawings to be made hereafter. Our premises here are that the Earth, Mars, Jupiter, and Saturn are a random sample of a natural class of major planets, — a class which, though (so far as we know) it is very small, yet *may* be very extensive, comprising whatever there may be that revolves in a circular orbit around a great sun, is nearly spherical, shines with reflected light, is very large, etc. Now the examples of major planets that we can examine all rotate on their axes; whence we suppose that Mercury, Venus, Uranus, and Neptune, since they possess, so far as we know, all the properties common to the natural class to which the Earth, Mars, Jupiter, and Saturn belong, possess this property likewise. The points to be observed are, first, that any small class of things may be regarded as a mere sample of an actual or possible large class having the same properties and subject to the same conditions; second, that while we do not know what all these properties and conditions are, we do know some of them, which some may be considered as a random sample of all; third, that a random selection without replacement from a small class may be regarded as a true random selection from that infinite class of which the finite class is a random selection. The formula of the analogical inference presents, therefore, three premises, thus: —

S', S'', S''' are a random sample of some undefined class X, of whose characters P', P'', P''' are samples.

Q is P', P'', P'''.

S', S'', S''', are R's.

Hence, Q is an R.

We have evidently here an induction and an hypothesis followed by a deduction; thus,—

Every X is, for example, P', P'', P''', etc.	S', S'', S''', etc., are samples of the X's.
Q is found to be P', P'', P''', etc.	S', S'', S''', etc., are found to be R's.
Hence, hypothetically, Q is an X.	Hence, inductively, every X is an R.

Hence, deductively, Q is an R.*

An argument from analogy may be strengthened by the addition of instance after instance to the premises, until it loses its ampliative character by the exhaustion of the class and becomes a mere deduction of that kind called *complete induction*, in which, however, some shadow

* That this is really a correct analysis of the reasoning can be shown by the theory of probabilities. For the expression

$$\frac{(p+q)!}{p!\,q!}\cdot\frac{(\pi+\rho)!}{\pi!\,\rho!}\cdot\frac{(p+\pi)!\,(q+\rho)!}{(p+\pi+q+\rho)!}$$

expresses at once the probability of two events; namely, it expresses first the probability that of $p+q$ objects drawn without replacement from a lot consisting of $p+\pi$ objects having the character R together with $q+\rho$ not having this character, the number of those drawn having this character will be p; and second, the same expression denotes the probability that if among $p+\pi+q+\rho$ objects drawn at random from an infinite class (containing no matter what proportion of R's to non-R's), it happens that $p+\pi$ have the character R, then among any $p+q$ of them, designated at random, p will have the same character. Thus we see that the chances in reference to drawing without replacement from a finite class are precisely the same as those in reference to a class which has been drawn at random from an infinite class.

of the inductive character remains, as this name implies.

VIII.

Take any human being, at random,—say Queen Elizabeth. Now a little more than half of all the human beings who have ever existed have been males; but it does not follow that it is a little more likely than not that Queen Elizabeth was a male, since we know she was a woman. Nor, if we had selected Julius Cæsar, would it be only a little more likely than not that he was a male. It is true that if we were to go on drawing at random an indefinite number of instances of human beings, a slight excess over one-half would be males. But that which constitutes the probability of an inference is the proportion of true conclusions among all those which could be derived *from the same precept.* Now a precept of inference, being a rule which the mind is to follow, changes its character and becomes different when the case presented to the mind is essentially different. When, knowing that the proportion r of all M's are P's, I draw an instance, S, of an M, without any other knowledge of whether it is a P or not, and infer with probability, r, that it is P, the case presented to my mind is very different from what it is if I have such other knowledge. In short, I cannot make a valid probable inference without taking into account whatever knowledge I have (or, at least, whatever occurs to my mind) that bears upon the question.

The same principle may be applied to the statistical deduction of Form IV. If the major premise, that the proportion r of the M's are P's, be laid down first, before the instances of Ms are drawn, we really draw our inference concerning those instances (that the proportion r of them will be P's) in advance of the drawing, and therefore before we know whether they are P's or not. But if we draw the instances of the M's first, and after the examination of them decide what we will select for the predicate of our major premise, the inference will generally be completely fallacious. In short, we have the rule that the major term P must be decided upon in advance of the examination of the sample; and in like manner in Form IV. (*bis*) the minor term S must be decided upon in advance of the drawing.

The same rule follows us into the logic of induction and hypothesis. If in sampling any class, say the M's, we first decide what the character P is for which we propose to sample that class, and also how many instances we propose to draw, our inference is really made before these latter are drawn, that the proportion of P's in the whole class is probably about the same as among the instances that are to be drawn, and the only thing we have to do is to draw them and observe the ratio. But suppose we were to draw our inferences without the predesignation of the character P; then we might in every case find some recondite character in which those instances would all agree. That, by the exercise of sufficient ingenuity, we should be sure to be able to do this, even if not a single other object of the class M possessed that character, is a matter of demonstration. For in geometry a curve may be drawn through any given series of points, without passing through any one of another given series of points, and this irrespective of the number of dimensions. Now, all the qualities of objects may be conceived to result from variations of a number of continuous variables; hence any lot of objects possesses some character in common, not possessed by any other. It is true that if the universe of quality is limited, this is not altogether true; but it remains true that unless we have some special premise from which to infer the contrary, it always *may* be possible to assign some common character of the instances S', S'', S''', etc., drawn at random from among the M's, which does not belong to the M's generally. So that if the character P were not predesignate, the deduction of which our induction is the apagogical inversion would not be valid; that is to say, we could not reason that if the M's did not generally possess the character P, it would not be likely that the S's should all possess this character.

I take from a biographical dictionary the first five names of poets, with their ages at death. They are,

Aagard,	died	at	48.
Abeille,	"	"	76.
Abulola,	"	"	84.
Abunowas,	"	"	48.
Accords,	"	"	45.

These five ages have the following characters in common:—

1. The difference of the two digits composing the number, divided by three, leaves a remainder of *one.*

2. The first digit raised to the power indicated by the second, and then divided by three, leaves a remainder of *one.*

3. The sum of the prime factors of each age, including *one* as a prime factor, is divisible by *three.*

Yet there is not the smallest reason to believe that the next poet's age would possess these characters.

Here we have a *conditio sine quâ non* of valid induction which has been singularly overlooked by those who have treated of the logic of the subject, and is very frequently violated by those who draw inductions. So accomplished a reasoner as Dr. Lyon Playfair, for instance, has written a paper of which the following is an abstract. He first takes the specific gravities of the three allotropic forms of carbon, as follows: —

Diamond,	3.48
Graphite,	2.29
Charcoal,	1.88

He now seeks to find a uniformity connecting these three instances; and he discovers that the atomic weight of carbon, being 12,

$$\begin{aligned} \text{Sp. gr. diamond nearly} &= 3.46 = \sqrt[2]{12} \\ \text{" \; " \; graphite \; "} &= 2.29 = \sqrt[3]{12} \\ \text{" \; " \; charcoal \; "} &= 1.86 = \sqrt[4]{12} \end{aligned}$$

This, he thinks, renders it probable that the specific gravities of the allotropic forms of other elements would, if we knew them, be found to equal the different roots of their atomic weight. But so far, the character in which the instances agree not having been predesignated, the induction can serve only to suggest a question, and ought not to create any belief. To test the proposed law, he selects the instance of silicon, which like carbon exists in a diamond and in a graphitoidal condition. He finds for the specific gravities —

Diamond silicon,	2.47
Graphite silicon,	2.33.*

* The author ought to have noted that this number is open to some doubt, since the specific gravity of this form of silicon appears to vary largely. If a different value had suited the theory better, he might have been able to find reasons for preferring that other value. But I do not mean to imply that Dr. Playfair has not dealt with perfect fairness with his facts, except as to the fallacy which I point out.

Now, the atomic weight of silicon, that of carbon being 12, can only be taken as 28. But 2.47 does not approximate to any root of 28. It is, however, nearly the cube root of 14, ($\sqrt[3]{\frac{1}{2} \times 28} = 2.41$), while 2.33 is nearly the fourth root of 28 ($\sqrt[4]{28} = 2.30$). Dr. Playfair claims that silicon is an instance satisfying his formula. But in fact this instance requires the formula to be modified; and the modification not being predesignate, the instance cannot count. Boron also exists in a diamond and a graphitoidal form; and accordingly Dr. Playfair takes this as his next example. Its atomic weight is 10.9, and its specific gravity is 2.68; which is the square root of $\frac{2}{3} \times 10.9$. There seems to be here a further modification of the formula not predesignated, and therefore this instance can hardly be reckoned as confirmatory. The next instances which would occur to the mind of any chemist would be phosphorus and sulphur, which exist in familiarly known allotropic forms. Dr. Playfair admits that the specific gravities of phosphorus have no relations to its atomic weight at all analogous to those of carbon. The different forms of sulphur have nearly the same specific gravity, being approximately the fifth root of the atomic weight 32. Selenium also has two allotropic forms, whose specific gravities are 4.8 and 4.3; one of these follows the law, while the other does not. For tellurium the law fails altogether; but for bromine and iodine it holds. Thus the number of specific gravities for which the law was predesignate are 8; namely, 2 for phosphorus, 1 for sulphur, 2 for selenium, 1 for tellurium, 1 for bromine, and 1 for iodine. The law holds for 4 of these, and the proper inference is that about half the specific gravities of metalloids are roots of some simple ratio of their atomic weights.

Having thus determined this ratio, we proceed to

inquire whether an agreement half the time with the formula constitutes any special connection between the specific gravity and the atomic weight of a metalloid. As a test of this, let us arrange the elements in the order of their atomic weights, and compare the specific gravity of the first with the atomic weight of the last, that of the second with the atomic weight of the last but one, and so on. The atomic weights are —

Boron,	10.9	Tellurium,	128.1
Carbon,	12.0	Iodine,	126.9
Silicon,	28.0	Bromine,	80.0
Phosphorus,	31.0	Selenium,	79.1
	Sulphur,	32.	

There are three specific gravities given for carbon, and two each for silicon, phosphorus, and selenium. The question, therefore, is, whether of the fourteen specific gravities as many as seven are in Playfair's relation with the atomic weights, not of the same element, but of the one paired with it. Now, taking the original formula of Playfair we find

Sp. gr. boron	= 2.68	$\sqrt[5]{\text{Te}} = 2.64$
3^d Sp. gr. carbon	= 1.88	$\sqrt[8]{\text{I}} = 1.84$
2^d Sp. gr. carbon	= 2.29	$\sqrt[6]{\text{I}} = 2.24$
1st Sp. gr. phosphorus	= 1.83	$\sqrt[7]{\text{Se}} = 1.87$
2^d Sp. gr. phosphorus	= 2.10	$\sqrt[6]{\text{Se}} = 2.07$

or five such relations without counting that of sulphur to itself. Next, with the modification introduced by Playfair, we have

1st Sp. gr. silicon	= 2.47	$\sqrt[4]{\frac{1}{2} \times \text{Br}} = 2.51$
2^d Sp. gr. silicon	= 2.33	$\sqrt[6]{2 \times \text{Br}} = 2.33$
Sp. gr. iodine	= 4.95	$\sqrt[2]{2 \times \text{C}} = 4.90$
1st Sp. gr. carbon	= 3.48	$\sqrt[3]{\frac{1}{3} \times \text{I}} = 3.48$

It thus appears that there is no more frequent agreement with Playfair's proposed law than what is due to chance.[1]

Another example of this fallacy was "Bode's law" of the relative distances of the planets, which was shattered by the first discovery of a true planet after its enunciation. In fact, this false kind of induction is extremely common in science and in medicine.[2] In the case of hypothesis, the correct rule has often been laid down; namely, that a hypothesis can only be received upon the ground of its having been *verified* by successful *prediction.* The term *predesignation* used in this paper appears to be more exact, inasmuch as it is not at all requisite that the ratio ρ should be given in advance of the examination of the samples. Still, since ρ is equal to 1 in all ordinary hypotheses, there can be no doubt that the rule of prediction, so far as it goes, coincides with that here laid down.

We have now to consider an important modification of the rule. Suppose that, before sampling a class of objects, we have predesignated not a single character but n characters, for which we propose to examine the samples. This is equivalent to making n different inductions from the same instances. The probable error in this case is that error whose probability for a simple induction is only $(\frac{1}{2})^n$, and the theory of probabilities shows that it in-

[1] As the relations of the different powers of the specific gravity would be entirely different if any other substance than water were assumed as the standard, the law is antecedently in the highest degree improbable. This makes it likely that some fallacy was committed, but does not show what it was.

[2] The physicians seem to use the maxim that you cannot reason from *post hoc* to *propter hoc* to mean (rather obscurely) that cases must not be used to prove a proposition that has only been suggested by these cases themselves.

creases but slowly with n; in fact, for $n = 1000$ it is only about five times as great as for $n = 1$, so that with only 25 times as many instances the inference would be as secure for the former value of n as with the latter; with 100 times as many instances an induction in which $n =$ 10,000,000,000 would be equally secure. Now the whole universe of characters will never contain such a number as the last; and the same may be said of the universe of objects in the case of hypothesis. So that, without any voluntary predesignation, the limitation of our imagination and experience amounts to a predesignation far within those limits; and we thus see that if the number of instances be very great indeed, the failure to predesignate is not an important fault. Of characters at all striking, or of objects at all familiar, the number will seldom reach 1,000; and of very striking characters or very familiar objects the number is still less. So that if a large number of samples of a class are found to have some very striking character in common, or if a large number of characters of one object are found to be possessed by a very familiar object, we need not hesitate to infer, in the first case, that the same characters belong to the whole class, or, in the second case, that the two objects are practically identical; remembering only that the inference is less to be relied upon than it would be had a deliberate predesignation been made. This is no doubt the precise significance of the rule sometimes laid down, that a hypothesis ought to be *simple*, — simple here being taken in the sense of familiar.

This modification of the rule shows that, even in the absence of voluntary predesignation, *some* slight weight is to be attached to an induction or hypothesis. And perhaps when the number of instances is not very small, it is enough to make it worth while to subject the inference to a regular test. But our natural tendency will be to attach too much importance to such suggestions, and we shall avoid waste of time in passing them by without notice until some stronger plausibility presents itself.

IX.

In almost every case in which we make an induction or a hypothesis, we have some knowledge which renders our conclusion antecedently likely or unlikely. The effect of such knowledge is very obvious, and needs no remark. But what also very often happens is that we have some knowledge, which, though not of itself bearing upon the conclusion of the scientific argument, yet serves to render our inference more or less probable, or even to alter the terms of it. Suppose, for example, that we antecedently know that all the *M*'s strongly resemble one another in regard to characters of a certain order. Then, if we find that a moderate number of *M*'s taken at random have a certain character, *P*, of that order, we shall attach a greater weight to the induction than we should do if we had not that antecedent knowledge. Thus, if we find that a certain sample of gold has a certain chemical character,—since we have very strong reason for thinking that all gold is alike in its chemical characters,—we shall have no hesitation in extending the proposition from the one sample to gold in general. Or if we know that among a certain people,—say the Icelanders,—an extreme uniformity prevails in regard to all their ideas, then, if we find that two or three individuals taken at random from among them have all any particular superstition, we shall be the more ready to infer that it belongs to the whole people from what we know of their uniformity. The influence of this sort of uniformity upon inductive conclusions was strongly insisted upon by Philodemus, and some very exact conceptions in regard to it may be gathered from the writings of Mr. Galton. Again, suppose we know of a certain character, P, that in whatever classes of a certain description it is found at all, to those it usually belongs as a universal character; then any induction which goes toward showing that all the M's are P will be greatly strengthened. Thus it is enough to find that two or three individuals taken at random from a genus of animals have three toes on each foot, to prove that the same is true of the whole genus; for we know that this is a *generic* character. On the other hand, we shall be slow to infer that all the animals of a genus have the same color, because color varies in almost every genus. This kind of uniformity seemed to J. S. Mill to have so controlling an influence upon inductions, that he has taken it as the centre of his whole theory of the subject.

Analogous considerations modify our hypothetic inferences. The sight of two or three words will be sufficient to convince me that a certain manuscript was written by myself, because I know a certain look is peculiar to it. So an analytical chemist, who wishes to know whether a solution contains gold, will be completely satisfied if it gives a precipitate of the purple of cassius with chloride of tin; because this proves that either gold or some hitherto unknown substance is present. These are examples of characteristic tests. Again, we may know of a certain person, that whatever opinions he holds he carries out with uncompromising rigor to their utmost logical consequences; then, if we find his views bear some of the marks of any ultra school of thought, we shall readily conclude that he fully adheres to that school.

There are thus four different kinds of uniformity and

non-uniformity which may influence our ampliative inferences: —

1. The members of a class may present a greater or less general resemblance as regards a certain line of characters.

2. A character may have a greater or less tendency to be present or absent throughout the whole of whatever classes of certain kinds.

3. A certain set of characters may be more or less intimately connected, so as to be probably either present or absent together in certain kinds of objects.

4. An object may have more or less tendency to possess the whole of certain sets of characters when it possesses any of them.

A consideration of this sort may be so strong as to amount to demonstration of the conclusion. In this case, the inference is mere deduction, — that is, the application of a general rule already established. In other cases, the consideration of uniformities will not wholly destroy the inductive or hypothetic character of the inference, but will only strengthen or weaken it by the addition of a new argument of a deductive kind.

X.

We have thus seen how, in a general way, the processes of inductive and hypothetic inference are able to afford answers to our questions, though these may relate to matters beyond our immediate ken. In short, a theory of the logic of verification has been sketched out. This theory will have to meet the objections of two opposing schools of logic.

The first of these explains induction by what is called the doctrine of Inverse Probabilities, of which the following is an example: Suppose an ancient denizen of the Mediterranean coast, who had never heard of the tides, had wandered to the shore of the Atlantic Ocean, and there, on a certain number m of successive days had witnessed the rise of the sea. Then, says Quetelet, he would have been entitled to conclude that there was a probability equal to $\frac{m+1}{m+2}$ that the sea would rise on the next following day.[1] Putting $m=0$, it is seen that this view assumes that the probability of a totally unknown event is $\frac{1}{2}$; or that of all theories proposed for examination one half are true. In point of fact, we know that although theories are not proposed unless they present some decided plausibility, nothing like one half turn out to be true. But to apply correctly the doctrine of inverse probabilities, it is necessary to know the antecedent probability of the event whose probability is in question. Now, in pure hypothesis or induction, we know nothing of the conclusion antecedently to the inference in hand. Mere ignorance, however, cannot advance us toward any knowledge; therefore it is impossible that the theory of inverse probabilities should rightly give a value for the probability of a pure inductive or hypothetic conclusion. For it cannot do this without assigning an antecedent probability to this conclusion; so that if this antecedent probability represents mere ignorance (which never aids us), it cannot do it at all.

The principle which is usually assumed by those who seek to reduce inductive reasoning to a problem in inverse probabilities is, that if nothing whatever is known about the frequency of occurrence of an event, then any one frequency is as probable as any other. But Boole

[1] See Laplace, "Théorie Analitique des Probabilités," livre ii. chap. vi.

has shown that there is no reason whatever to prefer this assumption, to saying that any one "constitution of the universe" is as probable as any other. Suppose, for instance, there were four possible occasions upon which an event might occur. Then there would be 16 "constitutions of the universe," or possible distributions of occurrences and non-occurrences. They are shown in the following table, where *Y* stands for an occurrence and *N* for a non-occurrence.

4 occurrences.	*3 occurrences.*	*2 occurrences.*	*1 occurrence.*	*0 occurrence.*
YYYY	*YYYN*	*YYNN*	*YNNN*	*NNNN*
	YYNY	*YNYN*	*NYNN*	
	YNYY	*YNNY*	*NNYN*	
	NYYY	*NYYN*	*NNNY*	
		NYNY		
		NNYY		

It will be seen that different frequencies result some from more and some from fewer different "constitutions of the universe," so that it is a very different thing to assume that all frequencies are equally probable from what it is to assume that all constitutions of the universe are equally probable.

Boole says that one assumption is as good as the other. But I will go further, and say that the assumption that all constitutions of the universe are equally probable is far better than the assumption that all frequencies are equally probable. For the latter proposition, though it may be applied to any one unknown event, cannot be applied to all unknown events without inconsistency. Thus, suppose all frequencies of the event whose occurrence is represented by *Y* in the above table are equally probable. Then consider the event which consists in a *Y* following a *Y* or an *N* following an *N*. The possible ways in which *this* event may occur or not are shown in the following table: —

3 occurrences.	*2 occurrences.*	*1 occurrence.*	*0 occurrence.*
Y Y Y Y	*Y Y Y N*	*Y Y N Y*	*Y N Y N*
N N N N	*N N N Y*	*N N Y N*	*N Y N Y*
	Y Y N N	*Y N N Y*	
	N N Y Y	*N Y Y N*	
	N Y Y Y	*Y N Y Y*	
	Y N N N	*N Y N N*	

It will be found that assuming the different frequencies of the first event to be equally probable, those of this new event are not so, — the probability of three occurrences being half as large again as that of two, or one. On the other hand, if all constitutions of the universe are equally probable in the one case, they are so in the other; and this latter assumption, in regard to perfectly unknown events, never gives rise to any inconsistency.

Suppose, then, that we adopt the assumption that any one constitution of the universe is as probable as any other; how will the inductive inference then appear, considered as a problem in probabilities? The answer is extremely easy;[1] namely, the occurrences or non-occurrences of an event in the past in no way affect the probability of its occurrence in the future.

Boole frequently finds a problem in probabilities to be indeterminate. There are those to whom the idea of an unknown probability seems an absurdity. Probability, they say, measures the state of our knowledge, and ignorance is denoted by the probability $\frac{1}{2}$. But I apprehend that the expression "the probability of an event" is an incomplete one. A probability is a fraction whose

[1] See Boole, "Laws of Thought."

numerator is the frequency of a specific kind of event, while its denominator is the frequency of a genus embracing that species. Now the expression in question names the numerator of the fraction, but omits to name the denominator. There is a sense in which it is true that the probability of a perfectly unknown event is one half; namely, the assertion of its occurrence is the answer to a possible question answerable by "yes" or "no," and of all such questions just half the possible answers are true. But if attention be paid to the denominators of the fractions, it will be found that this value of $\frac{1}{2}$ is one of which no possible use can be made in the calculation of probabilities.

The theory here proposed does not assign any probability to the inductive or hypothetic conclusion, in the sense of undertaking to say how frequently *that conclusion* would be found true. It does not propose to look through all the possible universes, and say in what proportion of them a certain uniformity occurs; such a proceeding, were it possible, would be quite idle. The theory here presented only says how frequently, in this universe, the special form of induction or hypothesis would lead us right. The probability given by this theory is in every way different — in meaning, numerical value, and form — from that of those who would apply to ampliative inference the doctrine of inverse chances.

Other logicians hold that if inductive and hypothetic premises lead to true oftener than to false conclusions, it is only because the universe happens to have a certain constitution. Mill and his followers maintain that there is a general tendency toward uniformity in the universe, as well as special uniformities such as those which we have considered. The Abbé Gratry believes that the tendency toward the truth in induction is due to a miraculous intervention of Almighty God, whereby we are led to make such inductions as happen to be true, and are prevented from making those which are false. Others have supposed that there is a special adaptation of the mind to the universe, so that we are more apt to make true theories than we otherwise should be. Now, to say that a theory such as these is *necessary* to explaining the validity of induction and hypothesis is to say that these modes of inference are not in themselves valid, but that their conclusions are rendered probable by being probable deductive inferences from a suppressed (and originally unknown) premise. But I maintain that it has been shown that the modes of inference in question are necessarily valid, whatever the constitution of the universe, so long as it admits of the premises being true. Yet I am willing to concede, in order to concede as much as possible, that when a man draws instances at random, all that he knows is that he *tries* to follow a certain precept; so that the sampling process might be rendered generally fallacious by the existence of a mysterious and malign connection between the mind and the universe, such that the possession by an object of an *unperceived* character might influence the will toward choosing it or rejecting it. Such a circumstance would, however, be as fatal to deductive as to ampliative inference. Suppose, for example, that I were to enter a great hall where people were playing *rouge et noir* at many tables; and suppose that I knew that the red and black were turned up with equal frequency. Then, if I were to make a large number of mental bets with myself, at this table and at that, I might, by statistical deduction, expect to win about half of them, — precisely as I might expect, from the results of these samples, to infer by induction the probable ratio of frequency of the turnings of red and black in the long run, if I did not know it. But could some devil look at each card before it was turned, and then influence me mentally to bet upon it or to refrain therefrom, the observed ratio in the cases upon which I had bet might be quite different from the observed ratio in those cases upon which I had not bet. I grant, then, that even upon my theory some fact has to be supposed to make induction and hypothesis valid processes; namely, it is supposed that the supernal powers withhold their hands and let me alone, and that no mysterious uniformity or adaptation interferes with the action of chance. But then this negative fact supposed by my theory plays a totally different part from the facts supposed to be requisite by the logicians of whom I have been speaking. So far as facts like those they suppose can have any bearing, they serve as major premises from which the fact inferred by induction or hypothesis might be deduced; while the negative fact supposed by me is merely the denial of any major premise from which the falsity of the inductive or hypothetic conclusion could in general be deduced. Nor is it necessary to deny altogether the existence of mysterious influences adverse to the validity of the inductive and hypothetic processes. So long as their influence were not too overwhelming, the wonderful self-correcting nature of the ampliative inference would enable us, even if they did exist, to detect and make allowance for them.

Although the universe need have no peculiar constitution to render ampliative inference valid, yet it is worth while to inquire whether or not it has such a constitution; for if it has, that circumstance must have its effect upon all our inferences. It cannot any longer be denied that the human intellect is peculiarly adapted to the comprehension of the laws and facts of nature, or at least of some of them; and the effect of this adaptation upon our reasoning will be briefly considered in the next section. Of any miraculous interference by the higher powers, we know absolutely nothing; and it seems in the present state of science altogether improbable. The effect of a knowledge of special uniformities upon ampliative inferences has already been touched upon. That there is a general tendency toward uniformity in nature is not merely an unfounded, it is an absolutely absurd, idea in any other sense than that man is adapted to his surroundings. For the universe of marks is only limited by the limitation of human interests and powers of observation. Except for that limitation, every lot of objects in the universe would have (as I have elsewhere shown) some character in common and peculiar to it. Consequently, there is but one possible arrangement of characters among objects as they exist, and there is no room for a greater or less degree of uniformity in nature. If nature seems highly uniform to us, it is only because our powers are adapted to our desires.

XI.

The questions discussed in this essay relate to but a small part of the Logic of Scientific Investigation. Let us just glance at a few of the others.

Suppose a being from some remote part of the universe, where the conditions of existence are inconceivably different from ours, to be presented with a United States Census Report, — which is for us a mine of valuable inductions, so vast as almost to give that epithet a new signification. He begins, perhaps, by comparing the ratio of indebtedness to deaths by consumption in counties whose names begin with the different letters of the alphabet. It is safe to say that he would find the ratio everywhere

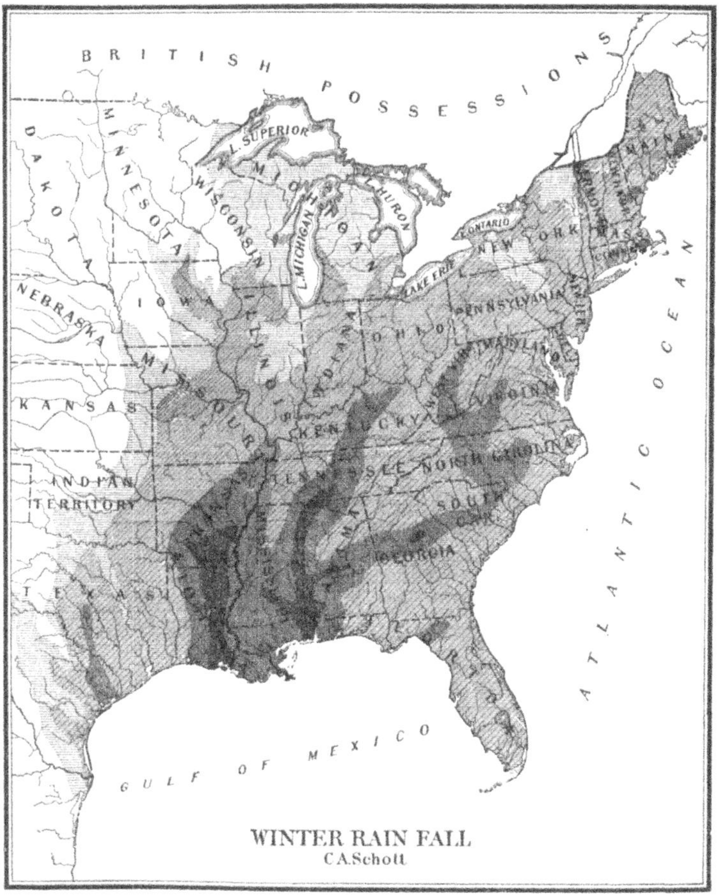

WINTER RAIN FALL
C.A.Schott

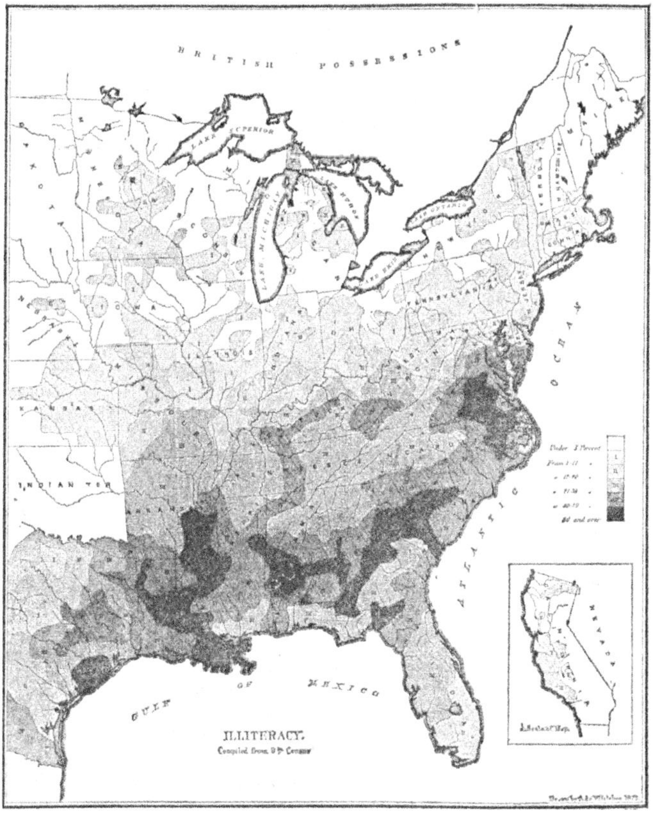

ILLITERACY.
Compiled from 9th Census

the same, and thus his inquiry would lead to nothing. For an induction is wholly unimportant unless the proportions of *P*'s among the *M*'s and among the non-*M*'s differ; and a hypothetic inference is unimportant unless it be found that *S* has either a greater or a less proportion of the characters of *M* than it has of other characters. The stranger to this planet might go on for some time asking inductive questions that the Census would faithfully answer, without learning anything except that certain conditions were independent of others. At length, it might occur to him to compare the January rain-fall with the illiteracy. What he would find is given in the following table[1]: —

REGION.	January Rain-fall.	Illiteracy.
	Inches.	Per cent.
Atlantic Sea-coast, Portland to Washington	0.92	11
Vermont, Northern and Western New York	0.78	7
Upper Mississippi River .	0.52	3
Ohio River Valley . . .	0.74	8
Lower Mississippi, Red River, and Kentucky	1.08	50
Mississippi Delta and Northern Gulf Coast	1.09	57
Southeastern Coast . . .	0.68	40

[1] The different regions with the January rain-fall are taken from Mr. Schott's work. The percentage of illiteracy is roughly estimated from the numbers given in the Report of the 1870 Census.

He would infer that in places that are drier in January there is, not always but generally, less illiteracy than in wetter places. A detailed comparison between Mr. Schott's map of the winter rain-fall with the map of illiteracy in the general census, would confirm the result that these two conditions have a partial connection. This is a very good example of an induction in which the proportion of *P*'s among the *M*'s is different, but not very different, from the proportion among the non-*M*'s. It is unsatisfactory; it provokes further inquiry; we desire to replace the *M* by some different class, so that the two proportions may be more widely separated. Now we, knowing as much as we do of the effects of winter rain-fall upon agriculture, upon wealth, etc., and of the causes of illiteracy, should come to such an inquiry furnished with a large number of appropriate conceptions; so that we should be able to ask intelligent questions not unlikely to furnish the desired key to the problem. But the strange being we have imagined could only make his inquiries hap-hazard, and could hardly hope ever to find the induction of which he was in search.

Nature is a far vaster and less clearly arranged repertory of facts than a census report; and if men had not come to it with special aptitudes for guessing right, it may well be doubted whether in the ten or twenty thousand years that they may have existed their greatest mind would have attained the amount of knowledge which is actually possessed by the lowest idiot. But, in point of fact, not man merely, but all animals derive by inheritance (presumably by natural selection) two classes of ideas which adapt them to their environment. In the first place, they all have from birth some notions, however crude and concrete, of force, matter, space, and time; and, in the next place, they have some notion of what sort of objects their fellow-beings are, and of how they will act on given occasions. Our innate mechanical ideas were so nearly correct that they needed but slight correction. The fundamental principles of statics were made out by Archimedes. Centuries later Galileo began to understand the laws of dynamics, which in our times have been at length, perhaps, completely mastered. The other physical sciences are the results of inquiry based on guesses suggested by the ideas of mechanics. The moral sciences, so far as they can be called sciences, are equally developed out of our instinctive ideas about human nature. Man has thus far not attained to any knowledge that is not in a wide sense either mechanical or anthropological in its nature, and it may be reasonably presumed that he never will.

Side by side, then, with the well established proposition that all knowledge is based on experience, and that science is only advanced by the experimental verifications of theories, we have to place this other equally important truth, that all human knowledge, up to the highest flights of science, is but the development of our inborn animal instincts.

NOTE A.

Boole, De Morgan, and their followers, frequently speak of a "limited universe of discourse" in logic. An unlimited universe would comprise the whole realm of the logically possible. In such a universe, every universal proposition, not tautologous, is false; every particular proposition, not absurd, is true. Our discourse seldom relates to this universe: we are either thinking of the physically possible, or of the historically existent, or of the world of some romance, or of some other limited universe.

But besides its universe of objects, our discourse also refers to a universe of characters. Thus, we might naturally say that virtue and an orange have nothing in common. It is true that the English word for each is spelt with six letters, but this is not one of the marks of the universe of our discourse.

A universe of things is unlimited in which every combination of characters, short of the whole universe of characters, occurs in some object. In like manner, the universe of characters is unlimited in case every aggregate of things short of the whole universe of things possesses in common one of the characters of the universe of characters. The conception of ordinary syllogistic is so unclear that it would hardly be accurate to say that it supposes an unlimited universe of characters;

but it comes nearer to that than to any other consistent view. The non-possession of any character is regarded as implying the possession of another character the negative of the first.

In our ordinary discourse, on the other hand, not only are both universes limited, but, further than that, we have nothing to do with individual objects nor simple marks; so that we have simply the two distinct universes of things and marks related to one another, in general, in a perfectly indeterminate manner. The consequence is, that a proposition concerning the relations of two groups of marks is not necessarily equivalent to any proposition concerning classes of things; so that the distinction between propositions in extension and propositions in comprehension is a real one, separating two kinds of facts, whereas in the view of ordinary syllogistic the distinction only relates to two modes of considering any fact. To say that every object of the class *S* is included among the class of *P*'s, of course must imply that every common character of the *P*'s is a common character of the *S*'s. But the converse implication is by no means necessary, except with an unlimited universe of marks. The reasonings in depth of which I have spoken, suppose, of course, the absence of any general regularity about the relations of marks and things.

I may mention here another respect in which this view differs from that of ordinary logic, although it is a point which has, so far as I am aware, no bearing upon the theory of probable inference. It is that under this view there are propositions of which the subject is a class of things, while the predicate is a group of marks. Of such propositions there are twelve species, distinct from one another in the sense that any fact capable of being expressed by a proposition of one of these species cannot be expressed by any proposition of another species. The following are examples of six of the twelve species: —

1. Every object of the class S possesses every character of the group π.

2. Some object of the class S possesses all characters of the group π.

3. Every character of the group π is possessed by some object of the class S.

4. Some character of the group π is possessed by all the objects of the class S.

5. Every object of the class S possesses some character of the group π.

6. Some object of the class S possesses some character of the group π.

The remaining six species of propositions are like the above, except that they speak of objects *wanting* characters instead of *possessing* characters.

But the varieties of proposition do not end here; for we may have, for example, such a form as this: "Some object of the class S possesses every character not wanting to any object of the class P." In short, the relative term "possessing as a character," or its negative, may enter into the proposition any number of times. We may term this number the *order* of the proposition.

An important characteristic of this kind of logic is the part that immediate inference plays in it. Thus, the proposition numbered 3, above, follows from No. 2, and No. 5 from No. 4. It will be observed that in both cases a universal proposition (or one that states the non-existence of something) follows from a particular proposition (or one that states the existence of something). All the immediate inferences are essentially of that nature. A particular proposition is never immediately inferable from a universal one. (It is true that from "no A exists" we can infer that "something not A exists;" but this is not properly an immediate inference, — it really supposes the additional premise that "something exists.") There are also immediate inferences raising and reducing the *order* of propositions. Thus, the proposition of the second order given in the last paragraph follows from "some S is a P." On the other hand, the inference holds, —

Some common character of the S's is wanting to everything except P's;

∴ Every S is a P.

The necessary and sufficient condition of the existence of a syllogistic conclusion from two premises is simple enough. There is a conclusion if, and only if, there is a middle term distributed in one premise and undistributed in the other. But the conclusion is of the kind called *spurious*[1] by De Morgan if, and only if, the middle term is affected by a "some" in both premises. For example, let the two premises be, —

Every object of the class S wants some character of the group μ;

Every object of the class P possesses some character not of the group μ.

The middle term μ is distributed in the second premise, but not in the first; so that a conclusion can be drawn. But, though both propositions are universal, μ is under a "some" in both; hence only a spurious conclusion can be drawn, and in point of fact we can infer both of the following: —

[1] On *spurious* propositions, see Mr. B. I. Gilman's paper in the *Johns Hopkins University Circular* for August, 1882. The number of such forms in any order is probably finite.

Every object of the class *S* wants a character other than some character common to the class *P*;

Every object of the class *P* possesses a character other than some character wanting to every object of the class *S*.

The order of the conclusion is always the sum of the orders of the premises; but to draw up a rule to determine precisely what the conclusion is, would be difficult. It would at the same time be useless, because the problem is extremely simple when considered in the light of the logic of relatives.

NOTE B.

A DUAL relative term, such as "lover," "benefactor," "servant," is a common name signifying a pair of objects. Of the two members of the pair, a determinate one is generally the first, and the other the second; so that if the order is reversed, the pair is not considered as remaining the same.

Let A, B, C, D, etc., be all the individual objects in the universe; then all the individual pairs may be arrayed in a block, thus:—

A : A	A : B	A : C	A : D	etc.
B : A	B : B	B : C	B : D	etc.
C : A	C : B	C : C	C : D	etc.
D : A	D : B	D : C	D : D	etc.
etc.	etc.	etc.	etc.	etc.

A general relative may be conceived as a logical aggregate of a number of such individual relatives. Let l denote "lover;" then we may write

$$l = \Sigma_i \Sigma_j (l)_{ij} (I : J)$$

where $(l)_{ij}$ is a numerical coefficient, whose value is 1 in case I is a lover of J, and 0 in the opposite case, and where the sums are to be taken for all individuals in the universe.

Every relative term has a negative (like any other term) which may be represented by drawing a straight line over the sign for the relative itself. The negative of a relative includes every pair that the latter excludes, and *vice versa*. Every relative has also a *converse*, produced by reversing the order of the members of the pair. Thus, the converse of "lover" is "loved." The converse may be represented by drawing a curved line over the sign for the relative, thus: $\breve{l}$. It is defined by the equation

$$(\breve{l})_{ij} = (l)_{ji}.$$

The following formulæ are obvious, but important:—

$$\bar{\bar{l}} = l \qquad \breve{\breve{l}} = l$$

$$\bar{\breve{l}} = \breve{\bar{l}}$$

$$(l -\!\!< b) = (\bar{b} -\!\!< \bar{l}) \qquad\qquad (l -\!\!< b) = (\breve{l} -\!\!< \breve{b}).$$

Relative terms can be aggregated and compounded like others. Using + for the sign of logical aggregation, and the comma for the sign of logical composition (Boole's multiplication, here to be called non-relative or internal multiplication), we have the definitions

$$(l + b)_{ij} = (l)_{ij} + (b)_{ij}$$

$$(l\,,\,b)_{ij} = (l)_{ij} \times (b)_{ij}.$$

The first of these equations, however, is to be understood in a peculiar way: namely, the + in the second member is not strictly addition, but an operation by which

$$0 + 0 = 0 \qquad 0 + 1 = 1 + 0 = 1 + 1 = 1.$$

Instead of $(l)_{ij} + (b)_{ij}$, we might with more accuracy write

$$0^{0^{(l)_{ij} + (b)_{ij}}}$$

The main formulæ of aggregation and composition are

$$\left\{\begin{array}{l}\text{If } l \prec s \text{ and } b \prec s, \text{ then } l + b \prec s. \\ \text{If } s \prec l \text{ and } s \prec b, \text{ then } s \prec l,b.\end{array}\right\}$$

$$\left\{\begin{array}{l}\text{If } l + b \prec s, \text{ then } l \prec s \text{ and } b \prec s. \\ \text{If } s \prec l,b, \text{ then } s \prec l \text{ and } s \prec b.\end{array}\right\}$$

$$\left\{\begin{array}{l}(l + b),s \prec l,s + b,s. \\ (l + s),(b + s) \prec l,b + s.\end{array}\right\}$$

The subsidiary formulæ need not be given, being the same as in non-relative logic.

We now come to the combination of relatives. Of these, we denote two by special symbols; namely, we write

lb for lover of a benefactor,

and

$l \dagger b$ for lover of everything but benefactors.

The former is called a particular combination, because it implies the *existence* of something *loved by* its relate and a *benefactor of* its correlate. The second combination is said to be *universal*, because it implies the *non-existence* of anything except what is either loved by its relate or a benefactor of its correlate. The combination lb is called a relative product, $l \dagger b$ a relative sum. The l and b are said to be undistributed in both, because if $l \prec s$, then $lb \prec sb$ and $l \dagger b \prec s \dagger b$; and if $b \prec s$, then $lb \prec ls$ and $l \dagger b \prec l \dagger s$.

The two combinations are defined by the equations

$$(lb)_{ij} = \Sigma_x (l)_{ix} (b)_{xj}$$

$$(l \dagger b)_{ij} = \Pi_x \{(l)_{ix} + (b)_{xj}\}$$

The sign of addition in the last formula has the same signification as in the equation defining non-relative multiplication.

Relative addition and multiplication are subject to the associative law. That is,

$$l \dagger (b \dagger s) = (l \dagger b) \dagger s,$$

$$l\,(bs) = (l\,b)\,s.$$

Two formulæ so constantly used that hardly anything can be done without them are

$$l\,(b \dagger s) \prec lb \dagger s,$$

$$(l \dagger b)\,s \prec l \dagger bs.$$

The former asserts that whatever is lover of an object that is benefactor of everything but a servant, stands to everything but servants in the relation of lover of a benefactor. The latter asserts that whatever stands to any servant in the relation of lover of everything but its benefactors, is a lover of everything but benefactors of servants. The following formulæ are obvious and trivial: —

$$ls + bs \prec (l + b)\,s$$

$$l,b \dagger s \prec (l \dagger s)\,,\,(b \dagger s).$$

Unobvious and important, however, are these: —

$$(l + b)\,s \prec ls + bs$$

$$(l \dagger s)\,,(b \dagger s) \prec l,b \dagger s.$$

There are a number of curious development formulæ. Such are

$$(l,b)\,s = \Pi_p\,\{l(s,p) + b\,(s,\bar{p})\}$$

$$l\,(b,s) = \Pi_p\{(l,p)\,b + (l,\bar{p})\,s\}$$

$$(l + b) \dagger s = \Sigma_p\,\{[l \dagger (s + p)],[b \dagger (s + \bar{p})]\}$$

$$l \dagger (b + s) = \Sigma_p\{[\,(l + p) \dagger b],[(l + \bar{p}) \dagger s]\}.$$

The summations and multiplications denoted by Σ and Π are to be taken non-relatively, and all relative terms are to be successively substituted for p.

The negatives of the combinations follow these rules:

$$\overline{l+b} = \bar{l},\bar{b} \qquad \overline{l,b} = \bar{l}+\bar{b}$$

$$\overline{l \dagger b} = \bar{l}\bar{b} \qquad \overline{lb} = \bar{l} \dagger \bar{b}$$

The converses of combinations are as follows: —

$$\widetilde{l+b} = \breve{l}+\breve{b} \qquad \widetilde{l,b} = \breve{l},\breve{b}$$

$$\widetilde{l \dagger b} = \breve{b} \dagger \breve{l} \qquad \widetilde{lb} = \breve{b}\breve{l}.$$

Individual dual relatives are of two types, —

A : A and A : B.

Relatives containing no pair of an object with itself are called *alio-relatives* as opposed to *self-relatives*. The negatives of alio-relatives pair every object with itself. Relatives containing no pair of an object with anything but itself are called *concurrents* as opposed to *opponents*. The negatives of concurrents pair every object with every other.

There is but one relative which pairs every object with itself and with every other. It is the aggregate of all pairs, and is denoted by ∞. It is translated into ordinary language by "coexistent with." Its negative is 0. There is but one relative which pairs every object with itself and none with any other. It is

$$(A : A) + (B : B) + (C : C) + \text{etc.};$$

is denoted by 1, and in ordinary language is "identical with —." Its negative, denoted by n, is "other than —," or "not."

No matter what relative term x may be, we have

$$0 \prec x \qquad x \prec \infty.$$

Hence, obviously

$$x + 0 = x \qquad x\,,\,\infty = x$$

$$x + \infty = \infty \qquad x\,,\,0 = 0.$$

The last formulæ hold for the relative operations; thus,

$$x \dagger \infty = \infty \qquad x0 = 0.$$

$$\infty \dagger x = \infty \qquad 0x = 0.$$

The formulæ

$$x + 0 = x \qquad x\,,\,\infty = x$$

also hold if we substitute the relative operations, and also 1 for ∞, and $\mathfrak{n}$ for 0; thus,

$$x \dagger \mathfrak{n} = x \qquad x1 = x.$$

$$\mathfrak{n} \dagger x = x \qquad 1x = x.$$

We have also

$$l + \bar{l} = \infty \qquad l\,,\,\bar{l} = 0.$$

To these partially correspond the following pair of highly important formulæ: —

$$1 \prec l \dagger \breve{\bar{l}} \qquad l\breve{\bar{l}} \prec \mathfrak{n}.$$

The logic of relatives is highly multiform; it is characterized by innumerable immediate inferences, and by various distinct conclusions from the same sets of premises. An example of the first character is afforded by Mr. Mitchell's F_{1v} following from $F_{1v'}$. As an instance of the second, take the premises,

Every man is a lover of an animal;

and

Every woman is a lover of a non-animal.

From these we can equally infer that

Every man is a lover of something which stands to each woman in the relation of not being the only thing loved by her, and that

Every woman is a lover of something which stands to each man in the relation of not being the only thing loved by him.

The effect of these peculiarities is that this algebra cannot be subjected to hard and fast rules like those of the Boolian calculus; and all that can be done in this place is to give a general idea of the way of working with it. The student must at the outset disabuse himself of the notion that the chief instruments of algebra are the inverse operations. General algebra hardly knows any inverse operations. When an inverse operation is identical with a direct operation with an inverse quantity (as subtraction is the addition of the negative, and as division is multiplication by the reciprocal), it is useful; otherwise it is almost always useless. In ordinary algebra, we speak of the "principal value" of the logarithm, etc., which is a direct operation substituted for an indefinitely ambiguous inverse operation. The elimination and transposition in this algebra really does depend, however, upon formulæ quite analogous to the

$$x + (-x) = 0 \qquad x \times \frac{1}{x} = 1,$$

of arithmetical algebra. These formulæ are

$$l, \bar{l} = 0 \qquad l\breve{\bar{l}} \prec \mathfrak{n}$$

$$l + \bar{l} = \infty \qquad 1 \prec l \dagger \breve{\bar{l}}.$$

For example, to eliminate s from the two propositions

$$1 \prec l\bar{s} \qquad 1 \prec \breve{s}b,$$

we relatively multiply them in such an order as to bring the two s's together, and then apply the second of the above formulæ, thus: —

$$1 \prec l\bar{s}\breve{s}b \prec l\mathfrak{n}b.$$

This example shows the use of the association formulæ in bringing letters together. Other formulæ of great importance for this purpose are

$$(b \dagger l)s \prec b \dagger ls \qquad b(l \dagger s) \prec bl \dagger s.$$

The distribution formulæ are also useful for this purpose.

When the letter to be eliminated has thus been replaced by one of the four relatives, — 0, ∞, 1, n, — the replacing relative can often be got rid of by means of one of the formulæ

$$l + 0 = l \qquad l, \infty = l$$

$$l \dagger \mathfrak{n} = \mathfrak{n} \dagger l = l \qquad l1 = 1l = l.$$

When we have only to deal with universal propositions, it will be found convenient so to transpose everything from subject to predicate as to make the subject 1. Thus, if we have given $l \prec b$, we may relatively add $\breve{\bar{l}}$ to both sides; whereupon we have

$$1 \prec l \dagger \breve{\bar{l}} \prec b \dagger \breve{\bar{l}}.$$

Every proposition will then be in one of the forms

$$1 \prec b \dagger l \qquad 1 \prec bl.$$

With a proposition of the form $1 \prec b \dagger l$, we have the right (1) to transpose the terms, and (2) to convert the terms. Thus, the following are equivalent: —

$$1 \prec b \dagger l$$

$$1 \prec l \dagger b \qquad 1 \prec \breve{b} \dagger \breve{l}$$

$$1 \prec \breve{l} \dagger \breve{b}.$$

With a proposition of the form $1 \prec bl$, we have only the right to convert the predicate giving $1 \prec \breve{l}\breve{b}$.

With three terms, there are four forms of universal propositions, namely: —

$$1 \prec l \dagger b \dagger s \quad 1 \prec l\,(b \dagger s) \quad 1 \prec lb \dagger s \quad 1 \prec lbs.$$

Of these, the third is an immediate inference from the second.

By way of illustration, we may work out the syllogisms whose premises are the propositions of the first order referred to in Note A. Let a and c be class terms, and let β be a group of characters. Let p be the relative "possessing as a character." The non-relative terms are to be treated as relatives, — a, for instance, being considered as "a coexistent with" and $\breve{a}$ as "coexistent with a that is." Then, the six forms of affirmative propositions of the first order are

$$1 \prec \breve{a} \dagger p \dagger \beta$$

$$1 \prec \breve{a}\,(p \dagger \beta) \qquad\qquad 1 \prec (\breve{a} \dagger p)\,\beta$$

$$1 \prec \breve{a}p \dagger \beta \qquad\qquad 1 \prec \breve{a} \dagger p\beta$$

$$1 \prec \breve{a}p\beta.$$

The various kinds of syllogism are as follows: —

1. *Premises:* $1 \prec \breve{a} \dagger p \dagger \beta \qquad 1 \prec \breve{c} \dagger p \dagger \bar{\beta}.$

Convert one of the premises and multiply,

$$1 \prec (\breve{a} \dagger p \dagger \beta)\,(\breve{\bar{\beta}} \dagger \breve{p} \dagger c) \prec \breve{a} \dagger p \dagger \beta\breve{\bar{\beta}} \dagger \breve{p} \dagger c$$
$$\prec \breve{a} \dagger p \dagger \mathfrak{n} \dagger \breve{p} \dagger c \prec \breve{a} \dagger p \dagger \breve{p} \dagger c.$$

The treatment would be the same if one or both of the premises were negative; that is, contained $\bar{p}$ in place of p.

2. *Premises:* $1 \prec \breve{a} \dagger p \dagger \beta \qquad 1 \prec \breve{c}\,(p \dagger \bar{\beta})$.

We have

$$1 \prec (\breve{a} \dagger p \dagger \beta)\,(\breve{\breve{\beta}} \dagger \breve{p})\,c \prec (\breve{a} \dagger p \dagger \breve{p})\,c.$$

The same with negatives.

3. *Premises:* $1 \prec \breve{a}\,(p \dagger \beta) \qquad 1 \prec \breve{c}\,(p \dagger \bar{\beta})$.

$$1 \prec \breve{a}\,(p \dagger \beta)\,(\breve{\breve{\beta}} \dagger \breve{p})\,c \prec \breve{a}\,(p \dagger \breve{p})\,c.$$

The same with negatives.

4. *Premises:* $1 \prec \breve{a} \dagger p \dagger \beta \qquad 1 \prec (\breve{c} \dagger p)\,\bar{\beta}$.

$$1 \prec (\breve{a} \dagger p \dagger \beta)\,\breve{\bar{\beta}}\,(\breve{p} \dagger c) \prec (\breve{a} \dagger p \dagger \beta\,\breve{\bar{\beta}})\,(\breve{p} \dagger c)$$
$$\prec (\breve{a} \dagger p)\,(\breve{p} \dagger c).$$

If one of the premises, say the first, were negative, we should obtain a similar conclusion,—

$$1 \prec (\breve{a} \dagger \bar{p})\,(\breve{p} \dagger c);$$

but from this again p could be eliminated, giving

$$1 \prec \breve{a} \dagger c, \quad \text{or} \quad \bar{a} \prec c.$$

5. *Premises:* $1 \prec \breve{a}\,(p \dagger \beta) \qquad 1 \prec (\breve{c} \dagger p)\,\bar{\beta}$.

$$1 \prec \breve{a}\,(p \dagger \beta)\,\breve{\bar{\beta}}\,(\breve{p} \dagger c) \prec \breve{a}p\,(\breve{p} \dagger c).$$

If either premise were negative, p could be eliminated, giving $1 \prec \breve{a}\,c$, or some a is c.

6. *Premises:* $1 \prec (\breve{a} \dagger p)\,\beta \qquad 1 \prec (\breve{c} \dagger p)\,\bar{\beta}$.

$$1 \prec (\breve{a} \dagger p)\,\beta\,\breve{\bar{\beta}}\,(\breve{p} \dagger c) \prec (\breve{a} \dagger p)\,\mathfrak{n}\,(\breve{p} \dagger c).$$

7. *Premises:* $1 \prec \breve{a} \dagger p \dagger \beta \qquad 1 \prec \breve{c}p \dagger \bar{\beta}$.

$$1 \prec (\breve{a} \dagger p \dagger \beta)\,(\breve{\bar{\beta}} \dagger \breve{p}c) \prec \breve{a} \dagger p \dagger \breve{p}c.$$

8. *Premises:* $1 \prec \breve{a}\,(p \dagger \beta) \qquad 1 \prec \breve{c}p \dagger \bar{\beta}$.

$$1 \prec \breve{a}\,(p \dagger \beta)\,(\breve{\bar{\beta}} \dagger \breve{p}c) \prec \breve{a}\,(p \dagger \breve{p}c).$$

9. *Premises:* $1 \prec (\breve{a} \dagger p)\,\beta \qquad 1 \prec \breve{c}p \dagger \bar{\beta}$.

$$1 \prec (\breve{a} \dagger p)\beta\,(\breve{\bar{\beta}} \dagger \breve{p}c) \prec (\breve{a} \dagger p)\,\breve{p}\,c.$$

If one premise is negative, we have the further conclusion $1 \prec \breve{a} c$.

10. *Premises:* $1 \prec \breve{a} p \dagger \beta \qquad 1 \prec \breve{c} p \dagger \bar{\beta}$.

$$1 \prec (\breve{a} p \dagger \beta)(\breve{\bar{\beta}} \dagger \breve{p} c) \prec \breve{a} p \dagger \breve{p} c.$$

11. *Premises:* $1 \prec \breve{a} \dagger p \dagger \beta \qquad 1 \prec \breve{c} \dagger p \bar{\beta}$.

$$1 \prec (\breve{a} \dagger p \dagger \beta)(\breve{\bar{\beta}} \breve{p} \dagger c) \prec (\breve{a} \dagger p) \breve{p} \dagger c.$$

We might also conclude

$$1 \prec \breve{a} \dagger p \dagger \mathfrak{n} \breve{p} \dagger c;$$

but this conclusion is an immediate inference from the other; for

$$(\breve{a} \dagger p) \breve{p} \dagger c \prec (\breve{a} \dagger p)(1 \dagger \mathfrak{n}) \breve{p} \dagger c \prec (\breve{a} \dagger p) 1 \dagger \mathfrak{n} \breve{p} \dagger c$$
$$\prec \breve{a} \dagger p \dagger \mathfrak{n} \breve{p} \dagger c.$$

If one premise is negative, we have the further conclusion $1 \prec \breve{a} \dagger c$.

12. *Premises:* $1 \prec \breve{a} (p \dagger \beta) \qquad 1 \prec \breve{c} \dagger p \bar{\beta}$.

$$1 \prec \breve{a} (p \dagger \beta)(\breve{\bar{\beta}} \breve{p} \dagger c) \prec \breve{a} (p \breve{p} \dagger c).$$

If one premise is negative, we have the further inference $1 \prec \breve{a} c$.

13. *Premises:* $1 \prec (\breve{a} \dagger p) \beta \qquad 1 \prec \breve{c} \dagger p \bar{\beta}$.

$$1 \prec (\breve{a} \dagger p) \beta (\breve{\bar{\beta}} \breve{p} \dagger c) \prec (\breve{a} \dagger p)(\mathfrak{n} \breve{p} \dagger c).$$

14. *Premises:* $1 \prec \breve{a} p \dagger \beta \qquad 1 \prec \breve{c} \dagger p \bar{\beta}$.

$$1 \prec (\breve{a} p \dagger \beta)(\breve{\bar{\beta}} \breve{p} \dagger c) \prec \breve{a} p \breve{p} \dagger c.$$

If one premise is negative, we have the further spurious inference $1 \prec \breve{a} \mathfrak{n} \dagger c$.

15. *Premises:* $1 \prec \breve{a} \dagger p \beta \qquad 1 \prec \breve{c} \dagger p \bar{\beta}$.

$$1 \prec (\breve{a} \dagger p \beta)(\breve{\bar{\beta}} \breve{p} \dagger c) \prec \breve{a} \dagger p (\mathfrak{n} \breve{p} \dagger c).$$

We can also infer $1 \prec (\breve{a} \dagger p \mathfrak{n}) \breve{p} \dagger c$.

16. *Premises:* $1 \prec \breve{a} \dagger p \dagger \beta \qquad 1 \prec \breve{c} p \bar{\beta}.$

$$1 \prec (\breve{a} \dagger p \dagger \beta) \breve{\bar{\beta}} \breve{p} c \prec (\breve{a} \dagger p) \breve{p} c.$$

If one premise is negative, we can further infer $1 \prec \breve{a} c$.

17. *Premises:* $1 \prec \breve{a} (p \dagger \beta) \qquad 1 \prec \breve{c} p \bar{\beta}.$

$$1 \prec \breve{a} (p \dagger \beta) \breve{\bar{\beta}} \breve{p} c \prec \breve{a} p \breve{p} c.$$

If one premise is negative, we have the further spurious conclusion $1 \prec \breve{a} \mathfrak{n} c$.

18. *Premises:* $1 \prec (\breve{a} \dagger p) \beta \qquad 1 \prec \breve{c} p \bar{\beta}.$

$$1 \prec (\breve{a} \dagger p) \beta \breve{\bar{\beta}} \breve{p} c \prec (\breve{a} \dagger p) \mathfrak{n} \breve{p} c.$$

19. *Premises:* $1 \prec \breve{a} p \dagger \beta \qquad 1 \prec \breve{c} p \bar{\beta}.$

$$1 \prec (\breve{a} p \dagger \beta) \breve{\bar{\beta}} \breve{p} c \prec \breve{a} p \breve{p} c.$$

If one premise is negative, we further conclude $1 \prec \breve{a} \mathfrak{n} c$.

20. *Premises:* $1 \prec \breve{a} \dagger p \beta \qquad 1 \prec \breve{c} p \bar{\beta}.$

$$1 \prec (\breve{a} \dagger p\beta) \breve{\bar{\beta}} \breve{p} c \prec (\breve{a} \dagger p \mathfrak{n}) \breve{p} c.$$

21. *Premises:* $1 \prec \breve{a} p \beta \qquad 1 \prec \breve{c} p \bar{\beta}.$

$$1 \prec \breve{a} p \beta \breve{\bar{\beta}} \breve{p} c \prec \breve{a} p \mathfrak{n} \breve{p} c.$$

When we have to do with particular propositions, we have the proposition $\infty \overline{\prec} 0$, or "something exists;" for every particular proposition implies this. Then every proposition can be put into one or other of the four forms

$$\infty \prec 0 \dagger l \dagger 0$$

$$\infty \prec (0 \dagger l) \infty$$

$$\infty \prec (0 \dagger l \infty$$

$$\infty \prec \infty l \infty.$$

Each of these propositions immediately follows from the one above it. The *enveloped* expressions which form the predicates have the remarkable property that each is either 0 or ∞. This fact gives extraordinary freedom in the use of the formulæ. In particular, since if anything not zero is included under such an expression, the whole universe is included, it will be quite unnecessary to write the $\infty \prec$ which begins every proposition.

Suppose that f and g are general relatives signifying relations of things to times. Then, Dr. Mitchell's six forms of two dimensional propositions appear thus: —

$$F_{11} = 0 \dagger f \dagger 0$$
$$F_{1v} = 0 \dagger f \, \infty$$
$$F_{u1} = \infty f \dagger 0$$
$$F_{1v'} = (0 \dagger f) \, \infty$$
$$F_{u'1} = \infty \, (f \dagger 0)$$
$$F_{uv} = \infty f \, \infty.$$

It is obvious that $l \dagger 0 \prec l$, for

$$l \dagger 0 \prec (l \dagger 0) \, \infty \prec l \dagger 0 \, \infty \prec l \dagger \mathfrak{n} \prec l.$$

If then we have $0 \dagger f \dagger 0$ as one premise, and the other contains g, we may substitute for g the product (f, g).

$$g \prec g, \infty \prec g, (0 \dagger f \dagger 0) \prec g, f.$$

From the two premises

$$\infty \, (f \dagger 0) \qquad \text{and} \qquad 0 \dagger g \, \infty,$$

by the application of the formulæ

$$ls, (b \dagger \bar{s}) \prec (l, b) \, s$$
$$sl, (\bar{s} \dagger b) \prec s \, (l, b),$$

we have

$$\{\infty \, (f \dagger 0)\}, (0 \dagger g \infty) \prec \infty \{(f \dagger 0), g \infty\} \prec \infty \, (f, g) \, \infty.$$

These formulæ give the first column of Dr. Mitchell's rule on page 90.

The following formulæ may also be applied: —

1. $(0 \dagger f \dagger 0),(0 \dagger g \dagger 0) \;=\; 0 \dagger (f,g) \dagger 0.$
2. $(0 \dagger f)\infty \quad (0 f \breve{g} \dagger 0) \mathrel{-\!\!<} (0 \dagger f)(\breve{g} \dagger 0).$
3. $(0 \dagger f)\infty \quad \infty(\breve{g} \dagger 0) \;=\; (0 \dagger f)(\breve{g} \dagger 0) + (0 \dagger f)\,\mathfrak{n}\,(\breve{g} \dagger 0).$
4. $(0 \dagger f)\infty \quad (0 \dagger \breve{g})\infty \mathrel{-\!\!<} (0 \dagger f)\breve{g}\infty.$
5. $(0 \dagger f \dagger 0)\,(0 \dagger \breve{g}\infty) \;=\; 0 \dagger (\breve{g}f, f) \dagger 0.$
6. $(0 \dagger f)\infty \quad (0 \dagger \breve{g}\infty) \;=\; (0 \dagger \breve{g}f, f)\infty.$
7. $(0 \dagger f)\infty,(0 \dagger g\infty) \;=\; (0 \dagger f, g\infty)\infty.$
8. $(0 \dagger f\infty)(0 \dagger g\infty) \;=\; 0 \dagger (f\breve{g}, \breve{g}f)\infty.$
9. $(0 \dagger f\infty),(0 \dagger g\infty) \;=\; 0 \dagger f\infty, g\infty.$
10. $(0 \dagger f \dagger 0)\infty\breve{g}\infty \;=\; 0 \dagger (f\breve{g}f, f) \dagger 0.$
11. $(0 \dagger f)\infty \quad \infty\breve{g}\infty \;=\; (0 \dagger f)\breve{g}\infty + (0 \dagger f)\,\mathfrak{n}\,\breve{g}\infty.$
12. $(0 \dagger f\infty) \quad \infty\breve{g}\infty \;=\; (0 \dagger f\breve{g}\infty) + (0 \dagger f\,\mathfrak{n}\,\breve{g}\infty).$
13. $\infty f\infty \quad \infty\breve{g}\infty \;=\; \infty f\breve{g}\infty + \infty f\,\mathfrak{n}\,\breve{g}\infty.$

When the relative and non-relative operations occur together, the rules of the calculus become pretty complicated. In these cases, as well as in such as involve *plural* relations (subsisting between three or more objects), it is often advantageous to recur to the numerical coefficients mentioned on page 187. Any proposition whatever is equivalent to saying that some complexus of aggregates[1] and products of such numerical coefficients is greater than zero. Thus,

$$\Sigma_i \Sigma_j l_{ij} > 0$$

means that something is a lover of something; and

$$\Pi_i \Sigma_j l_{ij} > 0$$

means that everything is a lover of something. We

[1] The sums of page 188.

shall, however, naturally omit, in writing the inequalities, the > 0 which terminates them all; and the above two propositions will appear as

$$\Sigma_i\Sigma_j l_{ij} \quad \text{and} \quad \Pi_i\Sigma_j l_{ij}.$$

The following are other examples: —

$$\Pi_i\Sigma_j(l)_{ij}(b)_{ij}$$

means that everything is at once a lover and a benefactor of something.

$$\Pi_i\Sigma_j(l)_{ij}(b)_{ji}$$

means that everything is a lover of a benefactor of itself.

$$\Sigma_i\Sigma_k\Pi_j(l_{ij} + b_{jk})$$

means that there is something which stands to something in the relation of loving everything except benefactors of it.

Let α denote the triple relative "accuser to — of —," and ε the triple relative "excuser to — of —. Then,

$$\Sigma_i\Pi_j\Sigma_k(\alpha)_{ijk}(\varepsilon)_{jki}$$

means that an individual i can be found, such, that taking any individual whatever, j, it will always be possible so to select a third individual, k, that i is an accuser to j of k, and j an excuser to k of i.

Let π denote "preferrer to — of —." Then,

$$\Pi_i\Sigma_j\Sigma_k(\alpha)_{ijk}(\varepsilon_{jki} + \pi_{kij})$$

means that, having taken any individual i whatever, it is always possible so to select two, j and k, that i is an accuser to j of k, and also is either excused by j to k or is something to which j is preferred by k.

When we have a number of premises expressed in this manner, the conclusion is readily deduced by the use of the following simple rules. In the first place, we have

$$\Sigma_i\Pi_j -\!\!< \Pi_j\Sigma_i.$$

In the second place, we have the formulæ

$$\{\Pi_i \varphi(i)\}\{\Pi_j \psi(j)\} = \Pi_i\{\varphi(i) \cdot \psi(i)\}.$$
$$\{\Pi_i \varphi(i)\}\{\Sigma_j \psi(j)\} -\!\!< \Sigma_i\{\varphi(i) \cdot \psi(i)\}.$$

In the third place, since the numerical coefficients are all either *zero* or *unity*, the Boolian calculus is applicable to them.

The following is one of the simplest possible examples. Required to eliminate *servant* from these two premises:

First premise. There is somebody who accuses everybody to everybody, unless the unaccused is loved by some person that is servant of all to whom he is not accused.

Second premise. There are two persons, the first of whom excuses everybody to everybody, unless the unexcused be benefited by, without the person to whom he is unexcused being a servant of, the second.

These premises may be written thus:

$$\Sigma_h \Pi_i \Sigma_j \Pi_k (\alpha_{hik} + s_{jk} l_{ji}).$$
$$\Sigma_u \Sigma_v \Pi_x \Pi_y (\varepsilon_{uyx} + \bar{s}_{yv} b_{vx}).$$

The second yields the immediate inference,

$$\Pi_x \Sigma_u \Pi_y \Sigma_v (\varepsilon_{uyx} + \bar{s}_{yv} b_{vx}).$$

Combining this with the first, we have

$$\Sigma_x \Sigma_u \Sigma_y \Sigma_v (\varepsilon_{uyx} + \bar{s}_{yv} b_{vx})(\alpha_{xuv} + s_{yv} l_{yu}).$$

Finally, applying the Boolian calculus, we deduce the desired conclusion

$$\Sigma_x \Sigma_u \Sigma_y \Sigma_v (\varepsilon_{uyx} \alpha_{xuv} + \varepsilon_{uyx} l_{yu} + \alpha_{xuv} b_{vx}).$$

The interpretation of this is that either there is somebody excused by a person to whom he accuses somebody, or somebody excuses somebody to his (the excuser's) lover, or somebody accuses his own benefactor.

The procedure may often be abbreviated by the use of operations intermediate between Π and Σ. Thus, we may use Π', Π'', etc. to mean the products for all individuals except one, except two, etc. Thus,

$$\Pi_i'\Pi_j''(l_{ij} + b_{ji})$$

will mean that every person except one is a lover of everybody except its benefactors, and at most two non-benefactors. In the same manner, Σ', Σ'', etc. will denote the sums of all products of two, of all products of three, etc. Thus,

$$\Sigma_i''(l_{ii})$$

will mean that there are at least three things in the universe that are lovers of themselves. It is plain that if $m < n$, we have

$$\Pi^m -\!\!< \Pi^n \qquad \Sigma^n -\!\!< \Sigma^m.$$

$$(\Pi_i^m \varphi i)(\Sigma_j^n \psi j) -\!\!< \Sigma_i^{n-m}(\varphi i \,.\, \psi i)$$

$$(\Pi_i^m \varphi i)\,(\Pi_j^n \psi j) -\!\!< \Pi_i^{m+n}(\varphi i \,.\, \psi i)$$

Mr. Schlötel has written to the London Mathematical Society, accusing me of having, in my *Algebra of Logic*, plagiarized from his writings. He has also written to me to inform me that he has read that Memoir with "heitere Ironie," and that Professor Drobisch, the Berlin Academy, and I constitute a "lederliche Kleeblatt," with many other things of the same sort. Up to the time of publishing my Memoir, I had never seen any of Mr. Schlötel's writings; I have since procured his *Logik*, and he has been so obliging as to send me two cuttings from his papers, thinking, apparently, that I might be curious to see the passages that I had appropriated. But having examined these productions, I find no thought in them that I ever did, or ever should be likely to put forth as my own.

THE END.

Made in the USA
Monee, IL
04 November 2026

67385018R00125